AUTHORITY IN ISLAM

MEHDI MOZAFFARI

AUTHORITY IN Islam

FROM MUHAMMAD TO KHOMEINI

M. E. SHARPE, INC.
ARMONK, NEW YORK
LONDON, ENGLAND

Available in the United Kingdom and Europe from M. E. Sharpe, Publishers, 3 Henrietta Street, London WC2E 8LU.

Translated from the French, *Pouvoir Islamique: Une tête et mille caprices*.

Translated by Michel Vale.

Published simultaneously as Vol. XVI, No. 4 of *International Journal of Politics*.

Library of Congress Cataloging-in-Publication Data

Mozaffari, Mehdi, 1939–
Authority in Islam.

Translation of: Pouvoir islamique.
1. Islam and state. 2. Islamic countries—Politics and government.
3. Islam—Doctrines. I. Title.
JC49.M6713 1987 322'.1'0917671 86-13070
ISBN 0-87332-388-2

Printed in the United States of America

Contents

Preface

This book is not an orientalist study nor a study in Islamology, but an essay in political science. There has indeed been a veritable flood of books published on Islam in recent years, especially in the aftermath of the events that led to the revolution in Iran. They may be divided essentially into two categories: those of a literary, philosophical, and theological nature, and those of a semi-journalistic nature. In this relative abundance, political science has remained largely on the sidelines—strange when one considers that Islam, compared with other religions, has profoundly political aspects. Evidently the political importance of Islam, at both the theoretical and practical levels, has not been sufficient to attract the attention of political scientists. There are few universities or academic institutions where Islam is regarded as an independent area of study, representing a specific body of thought and a specific political system. It has generally been relegated to departments of literature or theology, and treated as a historical curiosity, as something exotic. Nor is this regrettable discrimination exclusive to the non-Muslim world; the situation is curiously the same in the Muslim world as well.

It is time, therefore, to put an end to this deplorable state of affairs and begin to study Islam as, for example, theories of liberal democracy or of Marxism are studied. This should also help to do away with the anarchy that reigns in what goes by the name of "Islamic studies." Numerous publications deal with Islam without defining what Islam is. What exactly does Islamic power mean? What is there that is specifically "Islamic" about Muslim regimes that invoke Islam as their authority? What criteria should be used to measure the degree and quality of

The system of transliteration used in this book is, with few exceptions, the one adopted by The Royal Asiatic Society.

that specifically Islamic ingredient in the regimes in question? So long as no adequate answers are forthcoming, discourse shall continue to wander in the clouds as it does at present.

This book is intended to help dispel these ambiguities. A number of criteria have been defined to permit a classification of the various "Islamic" regimes in accordance with proposed models. The validity of these criteria and these models is, of course, open for discussion, but that such an undertaking is needed is, it would seem, uncontestable.

The models used are primarily of ostensive value, providing the reader with a set of standards, a key, as it were, for carrying the inquiry further, either deductively (*qiyās*) or by mounting an independent judgment (*ijtihād*).

To better assess the implications of some of the contemporary manifestations of Islam, these are placed in a historical perspective, for it would be illusory to believe that contemporary "Islamic" models can be understood without constant reference to the classical models, and the Medina model in particular. This of course entails referring to the basic texts: the Koran, Muslim political writings (both classical and modern), constitutional texts, and the official declarations of Muslim leaders. It is generally appreciated that the Text occupies an exalted place in a religious system; but its function becomes even more important in the case of a religion, in this case Islam, whose political ambitions have retained their full vitality.

The impact of the Text and the Word on the individual behavior of Muslims and on the social and political events of their societies has scarcely been investigated. Scholars rarely give due consideration to these two extremely important factors. The general tendency is to believe that everything essential can be understood and analyzed solely on the basis of a few statistical data or a reductionist theoretical framework. Thus, the Text and the Word are disregarded, whereas they should be considered as two social factors ranking at least on a par with economics or ongoing political activities. Such an omission can have grave consequences, leading to monumental misconceptions.

One misconception of this order, for example, was the American attitude toward the events that led to the fall of the Shah in Iran. The memoirs and documents that have been published on the Islamic revolution leave no doubt that the American administration had no clear idea of what was taking place in Iran—despite the United States's deep infiltration into that country, and the extremely sophisticated means of intelligence-gathering it had at its disposal. How can the American discomfiture before a situation of such extreme gravity be explained? At one point President Carter seemed to have had an inkling that despite the enormous quantity of information at hand, there was something lacking that prevented a correct analysis of the situation. The presidential advisers were incapable of providing him with any additional information, no doubt because what they were looking for was not to be found in their computers. William Sullivan, United States Ambassador at Tehran, a man known for his political and diplomatic experience, had submitted a report on the Iranian situation. An excerpt from the report, which Zbigniew Brzezinski quotes in his book *Power and Principle*, is not without interest. "Khomeini," Sullivan wrote, "would be likely to return to Iran as a consequence of a religious-military accommodation and . . . would play a 'Gandhi-like' role" (p. 368). Such was the American administration's perception of the Ayatollah's personality. Somewhat later, it was the turn of Carter himself to sing Khomeini's praises, calling him "peaceful" and "nonviolent." The American president was doubtless the first to realize the grave mistake in judgment he had committed. The taking of the American hostages at Tehran, which led to Carter's defeat in the next presidential election, must be seen as one of the direct consequences of the president's error.

There is therefore good cause to speculate on the cause of this error. What the Americans lacked was a broad understanding of the situation, and that was impossible to achieve without the "*textual*" dimension, so to speak. Neither Sullivan, Cyrus Vance, Brzezinski, nor the President had taken the trouble to read the book Khomeini had written in his name, *Ḥukūmat-é Islami* (Islamic Government), in which he spells out very clearly what he intended to do once he assumed power. It indicated clearly that Khomeini had nothing in common with Gandhi in either thought or style. Moreover, it is sufficient to read the Koran to know that Islam is not Hinduism.

But it was not only the Americans who neglected to read the Text and to take it seriously. Some of Khomeini's closest collaborators and protégés were guilty of the same omission. Five days after his flight from Iran, Bani Sadr, ex-president of Khomeini's republic, told me sadly that he had read the Ayatollah's book only very belatedly, and that the delay had cost him dearly. Bani Sadr was also to regret his tardy reading of the publications of his new allies, the Mujāhidīn-é Khalq, who were to become his adversaries.

It seems that men of politics have a tendency to throw themselves facilely into the riskiest of actions without having first acquired an adequate knowledge of the situation at hand. But who will repair the harm their actions do to whole peoples and societies?

It is for these reasons that the author has chosen to rely on the Text as a fact of sociopolitical reality.

While this book, the text of which was completed in 1984, does not claim to have exhausted the subject of Islamic power, it may contribute to the development of a new approach in Islamic studies from a political science perspective.

I would like to express my deep appreciation for the support and encouragement of the Danish Research Council for Humanities, especially to Kalus Ferdinand, the leader of their Contemporary Islam Project, and to the secretaries of the Institute of Political Science of Aarhus, who typed the manuscript. Thanks also to Michel Vale for a superb translation and to Leslie English for her patient and gracious assistance with the editing.

Aarhus University
Denmark

AUTHORITY IN ISLAM

Introduction

A bitter struggle, political and theoretical, that has been rending the hearts and minds of Muslims for fourteen centuries has lost none of its fire even today. Quite the contrary. At issue basically is the problem of power and, secondarily, the state. What is the Islamic conception of power? Does Islam have a general theory of the state? The question of the political system of Islam is posed in these terms. A renewed clarity on this issue is essential for an understanding of the problems posed by and to Islam today. But one distinction must be drawn at the very outset: the distinction, namely, between the "fact of the Koran" and the "fact of Islam."[1]

In principle these two "facts" are one. The Koran is the founding text, and Islam is therefore necessarily derivative of it. But realities are different. Islam—the Koran face-to-face with realities, transcending its original locus in space and time (the Medina community)—does not always reflect the Koran, and the Koran in turn does not always have a grasp of the real affairs of men even when men claim it as their authority.

This discrepancy between the Koran as a codified, ordained, and invariable text, and the ever-shifting flux of earthly existence is at the source of the division of Islam into its diverse and innumerable sects, schools, and doctrines.

The "Koranic fact" consists of the totality of statements revealed by Allāh to his Emissary. These statements were not codified, at least officially, until about twenty years after the emissary departed his earthly abode.

A singular fact about the Koran is that it is not a preconceived theoretical work. It is a book for action or, more exactly, it is a narrative of the acts and experiences of a specific group of men within a specific geographical setting at a specific moment of history. That experience unfolded with the aim of destroying the established order and replacing it with a new one, the *Medina community*. Twenty-three years were needed for its achievement, from the day when the Architect announced (*naba'*) his intention to commence the new project, to the day he proclaimed the completion of his works (June 8, 632). "Today I have perfected your religion and have handed over to you my entire worldly goods. For you I established Islam as religion . . ." (The Koran V, 3).

All there was to be done, created, imagined, or dreamed, was done in these twenty-three years. This prodigious Experience unfolded in an extraordinary ambience of creativity, combat, war, and action. The Koran is the record of this ambience. It conveys to us the inner core of the events that took place within the leading group (its rivalries, ambitions, pledges, sacrifices, merits, and weaknesses). The relations between the Leader and his Companions (*Ṣaḥāba*) and his politicomilitary general staff (the famous Council of Ten, *'Ashara Mubashshara*) are described, as are the relations of the Leader with his team of leaders and with the rest of the Believers (*mu'minūn*) and simple supporters and sympathizers (*muslimūn*). These groupings together constituted the Community, and as part of that Community and on its behalf they maintained relations with groups and individuals that were not part of it. Thus the Koran narrates, sometimes in considerable detail, the history of conflicts and alliances between the Community and the rest of humankind outside it, in particular the tribes, the Jews, the Christians, the hypocrites (*munāfiqūn*), and the "infidels" (*kuffār*). The verses of the Koran may indeed have "descended" from heaven one by one, but it was the cadence of human events and the situation at each given moment in time that set their tone, and determined their form and their content. The verses of the Koran are essentially a distillation of the evolution of the Community from its beginning to its end.

But let us return to the question of whether Islam as "Koranic fact" has a conception of power. A few points should first be established prefatory to their more detailed treatment later on. First, it would be rather implausible to think that the experience undertaken by the men of Faith to establish a new universal order could take place without an authority, a leader, and an organization. Through battles and wars, through diplomatic initiatives, and through political alliances, these men were able to establish a community. Each act along the way required reflection, planning, and organization (division of labor) before being carried out. There is no need to elaborate the obvious point that the art and techniques of politics were present at all levels of action, and hence that power—inseparable from politics—was exercised in all its dimensions.

Second, it should be noted that the Koran has established a number of principles concerning authority and power. We shall return to these later on, in more detail; but for the present, suffice it to say that the Koran makes a clear distinction between the seat of authority and the seat of power. The first belongs exclusively to Allāh: there is only one authority, the one and only Allāh (*Lā ilāha illallāh)*. The second is occupied by Muḥammad as Allah's Emissary, "*Rasūl Allāh.*" This power, delegated or conferred, is not autonomous by definition, but depends rather on the authority of Allāh, from whom it must receive legitimacy to be obeyed. These two principles constitute the declaration of faith (*shahāda*) of every Muslim. He who proclaims his *shahāda* at the same time declares himself ready to act to further the achievement of the goal of the Experience. Later, especially after the end of the first epoque of Islam, the *religious* character of the *shahāda* would come to predominate over its *political* character. There was no more experience to undertake; no more creativity, spontaneity, or innovative action, but a Form, a Framework, and a Dogma demanding adherence. Thus the "Koranic fact" came to an end, and the era of the "Islamic fact" began. As a manner of speaking we could say that the "Koranic fact" reached its culmination when the Emissary announced the achievement of his message. His death, which ensued shortly after that proclamation, completed the prophecy. There were no more verses to descend (*tanzīl*) from heaven, and no more revelations (*waḥy*). The concrete gave way to the abstract, and mystery took the place of clarity.

The group of companions, left without their Leader, could not believe that which was beyond belief. Moments of stupefaction, anguish, and groping ensued. But the *Ṣaḥāba* decided to continue the Adventure and maintain the astonishing dynamism of the message, and to do this they carried the Experience into other geographic regions, victory by victory. The conquered territories had to be administered, and so the Experience underwent a metamorphosis, and became Administration.

While he was alive Muḥammad avoided institutionalizing his works (or perhaps it was merely the lack of time that prevented him from doing so). The absence of institutional thinking was so marked that

Abū Bakr, his immediate successor, had no idea even what title he should use to make his decrees official. After many discussions and waverings, he finally opted, significantly enough, for the title of Caliphate (vicariate), but did not the "vicariate" have an air of the *provisional* and nonpermanent about it? Did not the title *Khalīfat al-Rasūl Allāh* signify that matters would continue as before and that continuity in change would be assured? This at least was what was hoped. Initially, the idea was not to institutionalize the caliphate, but reality took a different turn. In the endeavor to avoid institutionalization, the verses of the Koran would remain for some time in the same state as Muḥammad had left them: scattered, diffuse, and uncodified. Collations of them would begin under the first two caliphs, but would not be completed until the third, 'Uthmān b. 'Uffān (24–35/634-656).* One of these collations, that of Zayd b. Thābit, would be chosen by 'Uthmān as the authentic version and made official. The others, those ascribed by tradition to 'Alī, 'Ubay, or 'Abd Allāh Mas'ūd (who would be sentenced by 'Uthmān to flagellation because he refused to destroy his collation of the Koran) were eliminated.

Thus in but a short time the Medina Community had become no more than a memory. The power of the caliphate was established in other regions: Persia, Shām, Egypt. The vicariate gradually developed into a formidable institution with its administration, its laws, its jurists, and its police. The Koran, which in the meantime had become an eternally fixed and invariable codex, came to be used more and more as a means to justify actions undertaken by the caliph and his administration. Henceforth the way was clear for the transformation of the Experience into an institution and the institution into an empire—and even 'Uthmān's assassination failed to check its progress. It would perhaps be going too far to say that the codification of the message had a direct impact on the chain of events that ended with the great revolt (*Fitnat al-Kubrā*), but tradition has it that 'Uthmān was assassinated while working on his new codex, *Al-Qur'ān*.

The reign of 'Alī (35–40/656–661), the fourth caliph, was basically only a period of transition between the advent of the Message and its crystallization in the Empire beginning with the Umayyads. After the Umayyads, the Message—i.e., the "Koranic fact" that originally had contained nothing even vaguely reminiscent of the concept of a "state"—began increasingly to assume such features, acquiring the

*The first two dates refer to the Islamic calendar (anno Hejirae, A.H.) and the second two to the Christian calendar (anno Domini, A.D.).

qualities of monopoly and totality, and, most importantly, becoming secularized. Thereafter, drawing on the Koran was reserved for carefully and laboriously spelled-out conditions (*ṭahāra*). Reading the Koran became a ceremony and its meaning was considered beyond the grasp of the common man. To comprehend it, interpreters (*mufassirūn*), specialists (*fuqahā'*), and learned men (*'ulamā'*) were necessary. Creative innovation gave way to interpretation, and reflection to imitation (*taqlīd*). Those who came to be regarded as the truest successors of the prophet were those who proclaimed their intention to follow what Muḥammad had said or had done (*sunna*). Copying became the surest guarantee of authenticity. Mechanical rote stifled dynamic movement. The message of *Mecca* and *Medina* assumed the quality of *universality*, and the "fact of Islam" replaced the "fact of the Koran": the "Islamic state" was born. Whether it could have been otherwise is another question.

1. Approaches and Models

Approaches

The theoretical elaboration of the concept of "*Islamic state*" is of relatively recent date. To be sure, competition for power existed from the very outset, even during Muḥammad's lifetime. The numerous and bitter armed conflicts that set Muslim against Muslim testify to the passions political disputes can generate—from the division of Muslims within the Medina community (i.e., under the Prophet himself, among the different subgroups, most notably the "Emigrants" [*Muhājirūn*] and the "Helpers" [*Anṣār*]) to the war of Ṣiffīn between 'Alī and Mu'āwiya, and later the revolt of Ḥusayn son of 'Alī against Yazīd son of Mu'āwiya, and the fall of the Umayyads and their replacement by the 'Abbāsids.

In fact, it was these political schisms and struggles that subsequently gave rise to theoretical schisms into different sects, schools, and doctrines. The positions of the most important sects (Sunnī, Shī'ī, and Khārijī) will be discussed in a later chapter; for the moment let us examine briefly the different theoretical approaches to power and the state in Islam. Among the "classic" approaches we have necessarily had to be selective, yet our choices may be fairly said to be among the most representative.

Disregarding the approaches of *moralist* leanings, we may distinguish, broadly speaking, four fundamental political approaches[2] to the study of power and the state in Islam's classical period: philosophical, administrative, dogmatic, and sociological.

The philosophical approach

The most celebrated representative of the philosophical approach is unquestionably Al-Fārābī (259–339/872–950), who lived during the apogee of Hellenism in Islam. He himself was a Muslim Hellenist *par excellence* and in fact also bore the appelation *Magistrat Secundus*. His political thought is basically organic and utopian: *organic* in the sense that the state (the "city," or *medina*, to use his own term) is conceived of as a living body composed of organs, the importance of which varies depending on their function. It is a hierarchically organized body with interdependent members. "The dominant organ is by nature the most perfect and the most complete unto itself in all that is proper to it, in that it possesses to the utmost degree all that of which every other organ partakes, and in that below it exist organs that dominate other subordinate organs."

For Fārābī, it is the Leader who is the cause[3] of the city and not vice versa.

> Just as the heart is the first instance, and is the cause of the existence of all the other organs of the body, of their powers and of their hierarchy, such that if one of these organs is in distress it is the heart that provides the means to stem that distress, so (too) the Leader of the city must exist first, in order to be the cause of the existence of the city and of its parts, and the cause of . . . their hierarchy, such that if one of them is in distress it is he (the Leader) who can stem that distress.[4]

Fārābī's utopianism is manifest at two levels: in the qualities of the Leader of the virtuous city (*al-madinat al-faḍila*) and in the divisions of the different cities that he enumerates. Fārābi's concept of the Leader was deeply influenced by Plato, while his division of the city was influenced by Aristotle.

Inspired by Plato's *Republic* and influenced by the Shī'ī idea of the *imām*, Fārābī desired a leader who would be philosopher, prophet, *imām*, and king all at the same time. His view was, in essence, that a Leader is a man who has fulfilled his humanity to perfection, whose intelligence, having acquired all that was intelligible, had by that fact become intelligence *in actu* and intelligible *in actu*. It is to such a man that God makes his revelations, and he does so through the active intellect which lavishes these revelations on the acquired intellect of this man, whence they pass to passive intellect, and then to the imagination. Thus this man is, by reason of what he receives in his imagination, both sage and philosopher, proclaiming that which is and that which shall be in the most beautiful of languages.[5]

In saying this Fārābī is aware that the qualities he enumerates in detail (a total of thirteen) in his work have a utopian flavor, and hence he also admits a certain realism, indicating that

> if one sole individual combining all of these conditions in himself be not found, yet be it that two are found, the one a sage and the other offering the rest of these conditions, both will be leaders of the City. If these conditions are shared among a group of individuals, one possessing wisdom and (each of the others possessing one of the other conditions) and if (these individuals) agree among themselves, they will (all) be virtuous leaders.[6]

The distinctions drawn among the different cities show clearly that the author of the *Conciliation of Plato and Aristotle* was broadly influenced by the *Politics* of Aristotle. To the virtuous City Fārābī opposes the ignorant City, the immoral and fickle Cities, and the City that has lost its way, each divided in turn into other cities.

The administrative approach

For the definitive institutionalization of the caliphate a centralized hierarchized administration based on defined norms and precise rules was necessary. Al-Māwardī (364–450/974–1058) is a typical example of a Muslim theoretician who attempted to define the nature and function of administration in all its various aspects. Numerous others have of course dealt with the same problem,[7] but the *scope* and *rigor* of Māwardī's works place them in a category by themselves. His treatise on public law, *Constitutionis politicae* (al-Aḥkām al-sulṭāniyya), explores the caliphate (or imāmate) and the emirate, jurisprudence, the army, taxes, sanctions, legal punishment, etc. The vast domain over which his inquiry ranges is rendered with a mastery that, in Laoust's words, "only age can give." As for its rigor, he wisely avoids mixing good counsel (*naṣiḥa*) to the Prince, so much in fashion in the Muslim world, with the impersonal, technical side of administration.

The treatise gives the impression that for Māwardī, the institution of the caliphate was a thing of worth in itself, separate from and above the person of the caliph himself. He attaches, in fact, a particular importance to the *impersonal* quality of administration. It is the continuity of the caliphate as an institution that counts above all else. His thoroughly administrative approach was doubtless tributary to his nuanced attitude in politics, and in one sense Māwardī may be said to represent the views of the establishment of his day. He is thus the spokesman of the caliphate bureaucracy, and in that capacity disposed to any compromise to ensure the perpetuity of the administration.

Māwardī's administrative approach was clearly a response to the needs of the times. The 'Abbāsid caliphate was beset by serious problems caused essentially by the innumerable emirs, sultans, and princes who had established themselves on the territories belonging to the Baghdad caliphate. Power so diffuse and fragmented contained the inherent risk of bringing about the end of the caliphate as an institution. It was therefore imperative for the establishment to react to save the institution on which its own survival depended. Moreover, in the absence of a powerful caliphate, the administrative apparatus offered the only means to integrate, or at least tie together, so many disparate elements into a single body.

A rigid administration would hardly have been able to meet this need, and indeed would even have risked constricting that administration's sphere of influence and control even further. A more flexible administrative system was necessary, and it was to this project that Māwardī turned.

He succeeded brilliantly. Using every juridical expedient and formal device at his disposal, he deftly fashioned just the flexible, cohesive, and integrative administrative system that was needed. He drew first on "the Koran and Tradition (*sunna*)" and then on *consensus omnium* (*ijmā'*) and analogical arguments (*qiyās*). But he also made a broad appeal, in the name of public interest (*maṣlaḥa*), to other sources: custom (*'urf*), administrative jurisprudence (*'amal*), and the imperatives of the moment (*ḥukm al-waqt*). Thus a religious policy (*siyāsa*) was evolved, solicitous of protecting the legitimacy of the caliphate yet at the same time prudently making alternations and amendments (*istidrāk*) to existing institutions.[8]

The concern to protect the administration is so great in Māwardī that he even entertains the thought of a pluralist state instead of a unitary state. The caliphate is a unitary state when it has sufficient power to appoint emirs to govern the provinces who in their persons combine the requisite conditions and over whom the caliph can have relatively tight control. It is a pluralist state, more confederate than federal, when the caliph is obliged to work together with local dynasties, whose existence he accepts in return for their recognition of his preeminence.[9]

The dogmatic approach

The dogmatic approach may be regarded, broadly speaking, as a reaction to the two preceding ones. Fārābī's utopianism and Hellenism in general had no further theoretical ammunition to come to the aid of a caliphate in difficulty. The same, if to a lesser degree, applies to Māwardī's administrative approach; although it worked within the framework of the 'Abbāsid caliphate, it was unable to provide a theoretical basis the caliphate of Baghdad could use in its defense against the vehement attacks of Iranian and Egyptian Isma'ilism. In this situation, a return to dogma seemed to be the surest means for mounting a defense of the 'Abbāsid caliphate. Thus, the rationalist movement (mu'tazilism) was forced into retreat to make way for a dogmatism whose origins went back to the Ash'arī school, a bitter adversary of *mu'tazilism*. The most eminent representative of the new dogmatic wave was Al-Ghazālī (450–505/1058–1111) who was born the same year Māwardī died.

With Ghazālī, holder of the title *Ḥujjat ul-Islam*, we are already in the midst of the era of the "revival of religious sciences" (*Iḥyā' 'ulūm al-dīn*, title of one of Ghazālī's principal works). The choice of such a title is significant: if it was necessary to "revive" the religious sciences, by implication they had been dormant, if not dead.[10] Ghazālī

> noted with sadness and indignation that people in his time know much too little of religious law (*shar'*), not only in the cities but even more so in the villages and in the countryside. Therefore, a doctor of law (*faqīh*) officially charged with teaching religion must be placed in each village, tribe, or quarter.[11]

If the observation is true, ignorance of the *sharī'a* was widespread in Ghazālī's time at least among the common people, hence the necessity of a general remedy that consisted in a spectacular return to dogma, which entailed abandonment of philosophy and the condemnation of all nonconformist thought or attitude, specifically the *bāṭiniyya*.

Philosophy and heresy were two sides of the same coin, in Ghazālī's opinion. Whereas in his book *The Inconsistency of the Philosophers* (*Tahāfut al-Falāsafa*) he concentrates his attack on the philosophy of Fārābī and Avicenna (Ibn Sinā, 980–1037 A.D.), in the *Mustaẓhirī* he takes up the battle in earnest, saying that "the doctrine of the *bātiniyya*, seen from within, is that of the philosophers, and seen from without, that of the Shī'ī."[12]

But it should be borne in mind that Ghazālī, who ardently preached the dogmatization of the state, was also perfectly aware of the administrative problems besetting the caliphate, indeed, at least as aware of them as his illustrious predecessor Māwardī: while Māwardī had been exclusively in the service of the caliphate, Ghazālī was in practice simultaneously in the service of three different institutions. First, there was the *wizāra*, under the leadership of the powerful *wazir* (vizir) Niẓām al-Mulk, who raised his compatriot (both were from Tūs) from anonymity, charging him with the defense of the *shāfi'ī* credo against the *bāṭinī* criticisms. Through the *wizāra* Ghazālī would serve the Seljūq sultanate: first Malikshāh and then Sanjar. It was in honor of Sanjar that Ghazālī wrote his *Naṣihat al-mulūk* (Mirror of Princes). The caliphate was the third institution to benefit from Ghazālī's services. His book *Mustaẓhirī*, dedicated to the caliph Al-Mustaẓhir, was a treatise in defense of the institution of the caliphate.

Ghazālī's profound knowledge of the machinations and machinery of power at these three levels led him to the conclusion that Laoust summed up as follows:

> The strength that had enabled the prophet and the first caliphs to embark upon vast and bountiful conquests was lacking in the caliphate. The Turks had the strength, but lacked the convictions and religious training without which a military enterprise could only be a war of conquest and plunder. If strength is lacking, then "science," a missionary undertaking, a winning over of the leaders and the common people, is necessary—not of course the rational sciences, but the religious sciences such as tradition, Koranic exigesis, the *fiqh*, dogmatic theology, and the science of the heart.[13]

The sociological approach

The sociological approach is uncontestably the most "modern" of the four. Utopian constructions, administrative expedients, and dogmatic speculations vanish at a stroke in it. The state becomes a *human phenomenon* shorn of mystery and mystifications. The sociological approach, innovative and "revolutionary," which gave mankind (primarily non-Muslims) new instruments for scientific study and inquiry, was invented and developed by the unequalled Muslim theoretician, Ibn Khaldūn (734–808/1332–1406), who was also a preeminent political practitioner in every sense of the word. Statesman, man of intrigue, skillful mediator (Timūr, whom Ibn Khaldūn asked to spare the lives of the besieged inhabitants of Damas, was dazzled by his intelligence and ability), not to speak of his qualities as diplomat, judge, and teacher. Over and above his high posts and delicate missions, he was the man who "revolutionized" the social sciences. For Muslims, Ibn Khaldūn will always be the person who demystified the "Muslim state."

If one but considers the staggering role still played by mystification even today, six centuries after Ibn Khaldūn, at all levels of power in Muslim societies, the importance of Khaldūn's efforts will perhaps be better appreciated. He is the only Muslim thinker who ventured to reverse the sequence of phenomena in his analysis of power and the state. Instead of beginning at the top, the usual approach before and even after him, he commenced his analysis with a study of the concrete foundations of the state, moving progressively from there to its head:

> History studies human society (*al-ijtimā al-insānī*), i.e., universal civilization (*'umrān al-'ālam*). It deals with the nature (*ṭabī'a*) of this civilization, i.e., uncivilized life (*tawaḥḥush*) and social life (*ta'anus*), its singularities deriving from the clan and tribe (*'aṣabiyya*) and the ways by which one human group comes to dominate another. From the latter we are led to examine the birth of power (*mulk*), dynasties (*dawla*) and social classes (*marātib*).[14]

Clearly the state is merely an organization fashioned by men, with its origins hence on earth, and not in heaven. It has its causes and its roots in social life and in the way that life is lived, and accordingly this may produce an organization such as the state, or one of another type.[15] The state will take root not because God wills it, but "when the comfort and the luxury of the citizens drives out the wholesome primitivity of the bedouins." The wording alone, which reveals the materialist foundations of Khaldūn's approach, marked a revolution not only in Muslim thinking, but in the thought of the world at large.

Another of Ibn Khaldūn's "revolutionary" notions was his idea of *change* (*tabaddul*):

> The state of the world and of nations, their customs, and their beliefs do not keep the same form forever. They vary with time and pass through a succession of stages. So it is with persons, periods, and cities, and so it is as well with the regions of the world, with countries, with the times and with empires.

Applied to the political system and "beliefs" of Islam, the implication is that neither the caliphate, the imāmate, the *Umma* nor, if we may presume to be so bold, religions, are static and immutable institutions. Like any other phenomenon they too are subject to the "conditions" (*aḥwāl*) proper to nations and races," to material change; but "such change is like a latent illness not discovered until long after it has set in, and of which very few people are aware."[16]

Models

Islam has produced numerous models of power, and Muslim societies have at different points in history similarly produced their own versions of Islam, different from all the others. Of course, what this means is that Islam as such, general and unique in its kind, exists only in abstraction: nor does "true Islam" have any greater claim to existence, however vehement the claims or strong the avowals. The Truth of Islam is expressed through its concrete instances of both past and present. Thus, there are *Islamic truths*, just as there are *true, real, and one and only Islams*: indeed, one for each of us.

If the essence of Islam is in its existence—that is, if it lives and breathes only in particular times and in particular places—the various models of "Islamic power" must also be ruled by the same principles, relativity and plurality.

All models of Islam, therefore, are true and faithful models to the extent that they contain and reflect bits and pieces of Islamic truths. The totality of *these* truths constitutes Islamic truth. Other truths, i.e. *potential* truths, may, in any of their subsequent manifestations, confirm or invalidate present truths, but they will never leave them unchanged.

Of all the Islamic models, only one has been recognized by all Muslims at all times throughout history as the genuine model of Islamic government in the proper sense. This is, of course, the *Medina model* developed and administered by the Prophet himself. It sets the example, and has perennially inspired the realities and the dreams of all Muslims. In contrast, every model has been challenged by one or several groups of Muslims from its inception.

Beside the Medina model there was also the *first caliphate model* in four different forms: the elective caliphate (Abū Bakr), nominative caliphate ('Umar), deliberative caliphate ('Uthmān), and the insurrectional caliphate ('Alī).

Then there was the *monarchic caliphate model* of the Umayyads (661–750 A.D.), which a century later would give rise to the 'Abbāsid caliphate, with its strongly *imperial* qualities. The centralized empire of the ʿAbbāsids (750–1258 A.D.) gradually dissolved into a decentralized empire, embracing a multitude of more or less autonomous dynasties. As this was taking place, and as the star of the Fāṭimids (909–1171 A.D.) rose at Cairo, the caliphate lost its unitary character.

The Ottomans opened a new chapter in the history of Islam in which power acquired a different countenance. Colonization in turn again altered the general situation of the Muslim world, although the colonial presence alone was unable to effect any structural modifications in Muslim societies. The changes it did bring were always partial, superficial, and—significantly—always in accordance with the interests and needs of the colonizers.

Aside from the caliphate models, there were others that may aptly be called *alternative models*. The most important of these are Shiʿism and Khārijism. The latter succeeded in seizing power in a part of northern Africa, but its reign was of relatively short duration. Similarly, while Shiʿism, in the Fāṭimid variant, assumed power and entrenched itself in Egypt, and Zaydism, its other variant, reigned in Yemen until the *coup d'état* in 1961, the principal branch of Shiʿism, Imāmism, did not genuinely acquire power until February 1979 in Iran.

In our day there are political regimes that claim to be Islamic or that have installed themselves in societies that are predominantly Muslim. In either case, Islam brings the whole of its weight to bear on the life and institutions of these societies.

To say that Islam is a factor of political importance in a society where the majority of the population is Muslim is to state the obvious. But the weight it carries and the extent of its political influence varies from one society to another, and indeed, several political models may be differentiated in terms of this variation.

The cardinal question is how to measure the degree of "Islamicism" in each regime. What criteria should be applied to determine if power is totally Islamic, partially Islamic, or if it is based on a skillfully measured balance between Islam and secular thinking? In short, what are the constituent elements of the *basic Islamic personality* seen in its political dimension? At the individual level, to define the "basic postulates that over the centuries have shaped the cognitive scaffolding of Islamic consciousness"[17] is relatively easy. For example, Ḥasan Al-Bannā, founder of the Muslim Brotherhood, has produced a detailed list of these "basic postulates."[18] Difficulties are first encountered when one attempts to extend this basic Islamic personality into the political domain. Whereas the Substance of Islam (*Jawhar-ė Islam*) may be easy to define in general terms, this is far from being the case for its political substance. Muslims themselves are in disagreement over whether there is such a political core to Islam. Does Islam have a *political substance*? It is in such terms that the fundamental problem of power in Islam is posed.

Despite the weight it carries, the problem is a subordinate one relative to tangible realities. It matters very little in the final analysis whether Islam has a conception of power. The question then becomes scholastic. What is important, however, is that today, just as in the past, there are regimes, political organizations, and movements that hark back to Islam for their authority and in fact act as if Islam had effectively prescribed a political system with its own rules and norms. On the other hand, there are regimes in Muslim societies that have distanced themselves from this political Islam, or have tried to find a *modus vivendi* to coexist with Islam.

In all three cases, the key question is to determine what it is exactly that is the reference point of this distancing or rapprochement. In other words, what is the *specifically Islamic element* in a regime? In my view there is but one valid criterion that can bring any clarity into this question, namely *political shahāda*, i.e., the simultaneous acceptance of two things: first, that the seat of authority rests exclusively with Allāh, and second, that the condition of power is Muḥammad's prophecy. This means that where the latter is absent power is exercised by one of the following: the caliphate, the imāmate, or the *Umma*.

Application of this criterion should give us a rough measure of the degree of affinity or distance of any regime in any Muslim country *vis-à-vis* the political substance of Islam. It should also help us to arrive at a classification of the different Islamic models, or models of power effectively established in Muslim societies. Thus, one could say that if a regime solemnly recognizes the Authority of Allāh as unique and absolute, this regime represents a *monist model religious in essence* (e.g., Khomeini's regime in Iran).

If Allāh's authority is replaced by another authority, also absolute and unique, the regime remains monist, but its essence is other than Islamic. The Turkish and Afghan regimes are examples. While the former derives the whole of its legitimacy from the *nation*, the latter claims "proletarian" legitimacy.

Further, a regime may plainly tend toward a monist model, but because of the active presence of an "interfering" element it remains balancing, not daring to opt unequivocally for either side. This is the case with the Ba'th regimes in Iraq and Syria, Destour in Tunisia, and the FLN (Front de la Libération Nationale) regime in Algeria. Indeed the secularizing tendency of these regimes is as conspicuous as it is predominant. Nonetheless, none have been able to completely divorce themselves from the legitimating sources of Islam. We may therefore call them *semi-monist models*.

The third category comprises regimes or movements that, while recognizing Allāh as the seat of supreme authority, nonetheless interpret this authority for themselves. For them Allāh is politically equated with the *people* or the *nation*, so that in practical terms Allāh becomes one with the nation. We may accordingly call them *fusionist models*, the best examples of which are the Libyan regime and the movement of the Mujāhidīn-é Khalq in Iran.

Finally, a fourth category consists of *dualist models*, i.e., those that derive their legitimacy from two different sources: the seat of Islamic authority and that of another authority, e.g., the nation or the tribe. The 1906 Iranian Constitution, in which Allāh and the nation together constitute the two sources of legitimacy without the two being fused with one another, fits into this category. The same applies to the Egyptian regime, especially that of Sadat. The Saudi regime has the same features, except that the "nation" is replaced by the "tribe."

2. Authority, Legitimate Power, and Virtual Power

The analysis of any political system requires that a clear distinction be drawn between the seat of authority, the seat of legitimate power, and the seat of virtual power. Each of these covers different spheres, each with its own jurisdiction. They are ordered hierarchically, with authority at the top, and virtual power at the bottom of the scale, and legitimate power between the two. Authority is the supreme seat of legitimation. It is sovereign and independent. It is also the ultimate court of punishment and reward, the source of creation (legislation) and annihilation (annulment). While authority is the *right* to lead and command, power is the *supreme legitimate force*. Power obliges people to obey the will expressed by authority. Consequently, in its substance it is completely different from authority. In addition to power, there are in all societies *paralegitimate forces* (for example, in our era, political parties and trade unions). Although these forces can certainly bring their weight to bear, they are themselves unable to constitute legitimate power, and in this sense represent only *virtual power*. To these three seats we must add a fourth, *intermediate domain*, which in Islam is represented by the *'Ulamā'*.

Authors who have failed to draw this necessary distinction have found it difficult to place the sultanate and emirate in the classical political system of Islam; since in their thinking the caliphate constitutes authority in itself, the sultanate and emirate are by implication seen as the seat of power.[19]

There are three major objections to this approach: first, the caliphate is not an autonomous instance; it depends on the Prophecy, which itself is dependent on another instance, namely the authority of Allāh.

Second, it must be remembered that the sultanate and emirate were by no means institutions above the caliphate; they required the caliphate's agreement for their legitimacy, never vice-versa. Thus, they held sway in the domain of virtual power, not of effective power itself. Finally, the confusion of the seat of authority with the seat of power has generated misconceptions that have directly affected the analysis of contemporary Islam. Namely, the intense struggles that are currently taking place in Muslim societies most certainly do not have as their aim the conquest of the caliphate, which is nonexistent in any case. These struggles are centered on both authority and power, but the latter must in every case be legitimated by authority. For any legal power to be firm and lasting, such a legitimation is absolutely indispensable.

This being said, it should be observed that the supreme seat of legitimation is not the same for (to take an example) Kaddafi and Khomeini, just as it is different for the Muslim Brotherhood and the Mujāhidīn-é Khalq. Nevertheless, all four endlessly and fervidly quote Islam as their authority. Kaddafi and Khomeini both occupy the seat of "Islamic" power at present, but the authority to which they refer is not the same. Whereas Khomeini reigns in the name of Allāh and on behalf of the hidden Imām, Kaddafi believes in the possibility of a fusion between Allāh and the people. Thus, though they may appear identical, the two political systems are different. The same applies to the Muslim Brotherhood and the Mujāhidīn movement. The former feels closer to Khomeini, and in disagreement with the latter. The Mujāhidīn, in turn, having ideological affinities with the Libyan colonel, are in armed conflict with the Ayatollah.

These theoretical differences with regard to the seat of authority have serious political implications. Men, women, and children are inevitably implicated. The problem therefore is quite basic.

Extending our comparison to some other Muslim countries—Algeria, Turkey, and Saudi Arabia—we find that the differences separating them are even deeper and more clearly defined than those existing among the preceding examples. Although it does not fuse Allāh with the people, the Algerian regime places both in the seat of authority, albeit each with its own distinct carefully defined sphere. Turkey, in turn, removed Allāh from the seat of authority, and put the Nation in its place. Saudi Arabia represents another model. Not having a written constitution (with the exception of the Koran), the regime derives its legitimacy as much from its tribal structure as from the fact that it is Islamic.

These examples show that the choice of where to place the seat of authority can be of decisive importance for the nature of a regime and its political behavior.

The seat of authority

The confusion that has reigned among certain authors doubtless derives from the fact that Islam has not evolved a term of its own to specifically designate authority (*Auctoritas*). Various expressions such as *al-Amr*, *al-Ḥukm*, *al-Mulk* have been used to designate authority. The same words have been likewise applied to power (*Potestas*) and virtual power. The Koran, however, is very clear on the point that authority belongs exclusively to Allāh. It is he who

> created order: *khalq* and *amr*. God is in sovereign possession of a creative will (*irada khalqiyya* or *kwaniya*) and a legislative will (*irada tashri'yya*) . . . the Prophet himself, and all the more so a caliph, sultan, king, imām, master, father or husband can command only as a function of commands given by God.[20]

Thus, Allāh is the bedrock of any theoretical construction relating to the "Islamic state." Without the concept of Allāh, the political system of Islam would collapse into dust at the first blow. Allāh does not have solely a theological sense. In addition to his specifically Islamic qualities of the Unique, the Just, the Terrible, the Authoritative, the Clement, and so forth, he possesses a deeply political meaning and connotation. Political sovereignty belongs to him alone, wholly and totally. Nor does his transcendence alter anything in his immediate and political presence. Allāh is near man and "interposes between man and his own heart . . ." (VIII, 24).

The Koran materializes, concretizes, and politicizes Allāh, while at the same time rendering him inaccessible and untouchable. In this paradox (or miracle—*mu'jiza*) lies the source of the dynamism of the concept of Allāh, present and occult at the same time.

It is not surprising then that the undivided sovereignty of Allāh should be served by the Prophet as the first principle of adherence to Islam. By the act of *shahāda* (simultaneous witness and acceptance), consisting in *Lā ilāha illallāh* (no God but Allāh) the individual witnesses and accepts the absolute and unique authority of Allāh, otherwise beyond the individual's reach.

This monist vision has had a direct and immediate political purport since the moment it was proclaimed. At the time, the object was to destroy the prevailing political order (on the Arab peninsula) which was based on idolatry. Initially, the monist message of Muḥammad was unable to establish itself at Mecca, the prime site of revelation. The Messenger fled toward Medina (*hijra*) where the social climate was more favorable to monolithic ideas, doubtless because of the presence of Jewish populations. Thus, Muḥammad founded his Community at Medina, not at Mecca.

Later, after the death of the Prophet, the monist vision would have a major impact on the evolution of the Community. This monist vision was in fact the source of a unified vision of the community of men, the *Umma*. Conceived in this way, the community of believers is essentially a political community. More importantly perhaps, the political character of the *Umma* gives a religious connotation to the City which it is man's mission to construct on earth [21].

The monist vision has also contributed to the generation and/or fortification of personal power, authoritarian and despotic, all the while stressing power's mythical qualities in Muslim societies.

An examination of the concrete history of these societies is enough to confirm that for fourteen centuries there have always been people, tribes, families, groups, or castes who have reigned arbitrarily in various forms (caliphate, imamate, kingdom, empire, emirate), and indeed most often "*by delegation*," in the name of Allāh's sovereignty.

Although it may be true in theory that Allāh's authority is not transferable, this is not true in practice. Clearly, to give this authority a real meaning, it must assume some material form in the life of men. It must therefore have qualified executors, and accordingly it is at this level that the real problem of the seat of power and its occupancy arises. What type of men should be entrusted with translating Allāh's will into fact? This was, of course, the Prophet while he was alive; after him, the mission fell to his successors.

It follows that the seat of authority is not sufficient in itself in the real world. To function, the Islamic system needs other instances and other instruments. The executive instance (power) is that which must ensure the functioning of the system. But not in any way at all: it must do so in conformity with the design that the seat of authority has handed on to it. In Islam that design is the Koran.

The seat of power

At Mecca, Muḥammad proclaimed that authority belonged to Allāh the One. But the actual establishment of this authority in a unified community did not take place until later. For this to succeed, possession of power was required, and at the moment this eluded the Prophet. He gained that power later, not at Mecca but at Medina (*Yathrib*) where he founded his Community.

Thanks to the painstaking work of a number of specialists (R. B. Serjeant, R. Bell, A. F. L. Beeston, and M. J. Kister) and the aid of documents judged to be authentic, we are now able to retrace the process of the foundation of the Medina Community, the *Umma*.

Since this community is still considered by Muslims to be the perfect model of government, it is obviously essential to dwell upon it for a moment before comparing it with other "Islamic" models that came after it.

The Medina model

The tendency is generally to consider the *Umma* a specifically Islamic invention. This is not correct. It has been established that the *Umma*, both in its secular form, i.e., as a tribal confederation with a political character, and its theocratic form, based on a *dīn* (tribal religions and laws) existed before the advent of Islam. The existence of more than one *Umma* is in fact confirmed by the Koran itself. "Thus, we sent thee amongst an *Umma* before whom other *Umma*(s) have passed away . . ." (XIII, 29). In a Meccan sourate (XLIII, 22) addressed to Quraysh, the Koran qualifies the point further by saying:

> We never before thee sent a Warner to any Town (*qarya*), without the affluent saying: "We found our fathers following an *Umma*, and we follow their footsteps." He said: "And if I bring you better guidance than that followed by your fathers?" They said: "In the message ye are sent with, we are unbelievers." (XLIII, 23).

In concrete terms, this means that Muḥammad proposed to the Quraysh, from which he himself had come, a new *Umma* (theocratic confederation), claiming that it would be more beneficial for them than those to which their forefathers had belonged. What were the characteristics of this new and beneficial *Umma*? First, it was a unique and unified *Umma*, as the Koran tells us (XXI, 92). This implied the dissolving of other *Umma* and their integration into the new *Umma*. This new *Umma*, resting on secular foundations, must nevertheless be placed under the authority of a single God (Allāh), for a single *Umma* required a single God. "I am your Lord, so serve ye me" (XXI, 92). Consequently, the other gods (idolatry) must disappear, since their existence was incompatible with that of the single community. The third and last characteristic concerns the leadership and administration of the *Umma*. Who must lead it? Who had power? Before Islam, it was an Arab political tradition to rely on a judge or arbitrator (*ḥakam*). While leaving the foundations of this secular custom or rule intact, Muḥammad transformed it into a theocratic principle. He judged and arbitrated, not as a mere tribal chief, but initially in the quality of Warner (*Nabī*) (document A in the "Constitution" of Medina) and four or five years later as apostle (*Rasūl Allāh*) (document F).[22]

A study of the Constitution of Medina gives us a better idea of what actually took place during what was undoubtedly the most crucial period in the evolution of Islam and its material embodiment in the form of the new *Umma*.

Actually, that which is called the Constitution of Medina is a series of eight distinct texts, which are treaties of alliance (*ḥilf*) signed at Medina between the Prophet and his Medina partners. These treaties were written gradually over the course of the seven years following Muḥammad's arrival in Medina in 622.

The constituents

The first document constitutes the basic or fundamental law of the new *Umma*, who were the Muhājirūn (Emigrants), the Anṣār (Helpers), and the Yahūd (Hebrew tribes).

The Muhājirūn comprised the *mu'minūn* (sincere believers) and the *muslimūn* (converts) of the tribe of Quraysh, having left their tribe (*hājarū*) to join Muḥammad at Medina.

The distinction drawn between the *mu'minūn* and *muslimūn*, which exists in both the Medina documents and the Koran (XLIX, 14), has occasioned debate. R. B. Serjeant, referring to the ancient customs of the pre-Islamic Arab tribes, believes that the former were physically armed men of the tribes and that they in point of fact enjoyed a higher social status than the Muslims who came from the weaker (*ḍa'īf*) strata of the tribes. Whatever the case, if this distinction has any meaning, and there is reason to believe it does, it would follow that the believers—at least during the first years after the foundation of the *Umma*—were divided into two distinct groups, of higher and lower status.

The other question concerns the Muhājirūn themselves. In fact, not all the believers of the Quraysh were emigrants. There were also individuals who, while joining the new religion, maintained their tribal allegiance. These persons were not considered members of the *Umma*. Merely belonging to Islam was not sufficient; their acceptance into the *Umma* required their *hijra*, i.e., their total renunciation of tribal allegiance.

> Those who believe, who have gone forth and who have engaged in combat, with their possessions and their persons, on the path of Allāh, those who have given [them] refuge and succor, all these are taken into the brotherhood. With those who believe but have not gone forth, you shall have no affiliation until they go forth; if they ask for succor, in consideration of [their] religion, you must render it unto them except against tribes (*qawm*) with whom you maintain an alliance. Allāh is clairvoyant over all you do. (VIII, 72)

The Anṣār—the Medina tribes who gave refuge to the Muhājirūn and helped them establish themselves at Medina—were represented by the Aws and Khazraj, two powerful tribes, as well as by other tribes of lesser importance. The first document of the alliance was written in the house of Anas Ibn Mālik, a Khazraj from Banū Najjār.

The Hebrew tribes had settled in the Hijāz before the Christian era or, according to the author of *Kitāb al-Aghānī*,[23] after the Roman conquest of Palestine. At the founding of Islam, there were thirteen Hebrew tribes, of which only three would enter into direct conflict with the prophet. In the immediate pre-Islamic period the Jews did not have the forces to defend themselves without outside assistance (*dhallū wa-qualla' mtinā'uhum*). They therefore became clients or allies of the Aws and the Khazraj, who ensured their protection. By all appearances, it was through these two tribes that the Jews entered into the treaty of alliance. Ibn Isḥāq, the author of the *Sira*, mentions that the Jews participated in the treaty, not as disciples of a religion (*dīn*) but as members of specific tribes. Document A, §8 stipulates that "those Jews (*Yahūd* and not *al-Yahūd*) who are following us shall benefit by support and equality. It shall be forbidden to enter into an alliance against them":

وإنّه من تبعنا من يهودَ فإنّ له النصرَ والإُسوة، غيرِ مظلومين ولامتناصَرٍ عليهم

Consequently the Jews are full-fledged members of the *Umma*

وإنّ يهودَ بنى عَوْفٍ أمّةٌ مع المؤمنين، لليهود دينهم، وللمسلمين دينهم، مواليهم وأنفسهم

and are free to practice their own religion (doc. C, §2.a). Because they were full-fledged members, the Jews were to pay a *nafaqa*, as did Muslims (doc. C, §1, doc. F, §3, and doc. E, §3.b), and not *jizya*, the tax imposed on persons outside of the *Umma*.

The Umma

Thus, a new confederation was born on the ruins of preceding confederations. This new confederation would be unique and unified:

إنّهم أمّةٌ واحدة من دونِ الناس .

Despite its theocratic orientation, the new *Umma* was not ignorant of the political realities it had to confront. These were primarily of a tribal nature. The alliance took these realities into account and situated itself within the same overall framework that had existed even before Muḥammad; indeed, the texts of the alliance contain the same conventional phraseology that had been employed before Islam.

The various clauses of the alliance specify that the confederation must protect its members from outside attacks and the confederation declares itself responsible (in relation to them) for their blood spilt on the path of Allāh.

وإنّ المؤمنين يُبِئ بعضُهم على بعض بما نالَ دماءَهم فى سبيل الله .

Further, the confederation undertakes to conduct business in accordance with *ma'rūf* and *qist*. According to Serjeant, *ma'rūf*, in this textual context, signifies *shar'*, a word still used today in southwestern Arabia to designate the laws of custom. As for *qist*, according to *al-Nasāī* (*al-Sunan*, Cairo, 1349–1930, VIII, 18) it signifies *al-iṣlaḥ*, which Serjeant interprets as the equality of the blood price.

The power of Muḥammad

A study of the eight Medinan documents shows that the political power of Muḥammad was strengthened as the *Umma* became consolidated. In the first document (A), Muḥammad acts in his dual capacity of *Nabī* (Warner) and representative of the Muhājirūn. Apparently, the title of *Nabī* does not in itself give sufficient authority to impose one's will as judge and arbiter. At that time, its force derived essentially from the support provided by the Emigrants.

Shortly afterward (the year 1 of *hijir*), upon the death of the delegate (*naqīb*) of Banū Najjār, Muḥammad assumed the function of the deceased. The fact that he belonged to this tribe (on the maternal side) must surely have facilitated his taking this function (document C).

From the year 1 onward, according to document B, §4, Muḥammad was recognized as the supreme judge and arbiter of the *Umma*. Henceforth, any disputes that might arise within the confederation had to submit to the judgment and arbitration of Allāh (the seat of authority) and Muḥammad (the seat of power).

وإنّكم مهما اختلفتم فيه من شيء، فإنّ مَرَدّه إلى الله عزّ وجلّ، وإلى محمد صلى الله عليه وسلم.

Moreover, Muḥammad had the power to validate or invalidate any changes of alliance that might emerge among the tribes and their respective clients (doc. D, §3).

وإنّه لا يخرج منهم أحدٌ إلّا بإذن محمد صلى الله عليه وسلم.

The title of *Rasūl Allāh*, Apostle of God, appears for the first time only in document F, written in year 6 or 7. Muḥammad had already been recognized in document B as the supreme judge and arbiter of the Community, but without any mention of his title. Document F confirms his jurisdiction and his arbitration, and gives him this new and grandiose title.

فإنْ مَرَدَّه إلى الله عز وجلّ ، وإلى محمد رسول الله صلى الله عليه وسلم .

Thus, having gathered power in his hands, Muḥammad was henceforth able to lead the Community at will. As a symbol of his majesty (traditional), he forbade (according to a *hadīth* reported by Ibn Ḥanbal) taking his name in vain. In fact, in south Arabia, tradition forbade taking the name of sultans, saints, and all men of authority in vain.[24]

As a further mark of his growing power, Medina, his capital, was decreed a sacred enclave (*ḥaram*) in the Medinan Constitution.

Muḥammad, leader of the Community, was assisted in his functions by the Council of Ten (*'Ashara Mubashshara*). All the members of this council belonged to the tribe of Quraysh, but—significantly—to different clans: there were Abū Bakr and Ṭalḥa (clan of Taym), 'Umar and Sa'īd b. Zayd ('Adi), 'Abd al-Raḥmān b. 'Awf and Sa'd b. Abī Waqqās (Zahra), 'Ali (Hashīm), 'Uthman (Umayya), Al-Zubayr (Asad) and Abū 'Ubayda (Fahr). Each of these names represented a social and political force and was therefore able to effectively contribute to the establishment of a new power accorded to religious authority, as also defined by the Koran. The Council of Ten was with Muḥammad constantly; they lived near the Mosque of Medina and retained their political prerogatives after the death of the leader. Indeed, the four caliphs called Orthodox came from this council.''[25] Having discussed the *Umma* established at Medina, we may now determine the place of the various centers of power in the Medinan political system (see Figure 1).

The works of Muḥammad, that is, Islam, are regarded by Muslims as a miracle. In effect, Islam's ''miracle'' consists in its extraordinary art and capacity of transforming the secular into the theocratic, the specific into the universal, and the everyday into the eternal. Like any ambitious man of political calling, Muḥammad readily threw himself into the competitive fray, attaching a religious and transcendental quality to his role. He would enter into alliances with particular tribes, and the alliance would at once be transformed into a universal act. He would declaim on a question concerning everyday life, and his opinion assumed the dimensions of eternity. But these ''singularities'' were not unique to Muḥammad's power—any power, whatever its origin, is by its very nature generative, universal, and mythical. What differentiates Muḥammad, and thus undoubtedly distinguishes him from others, is

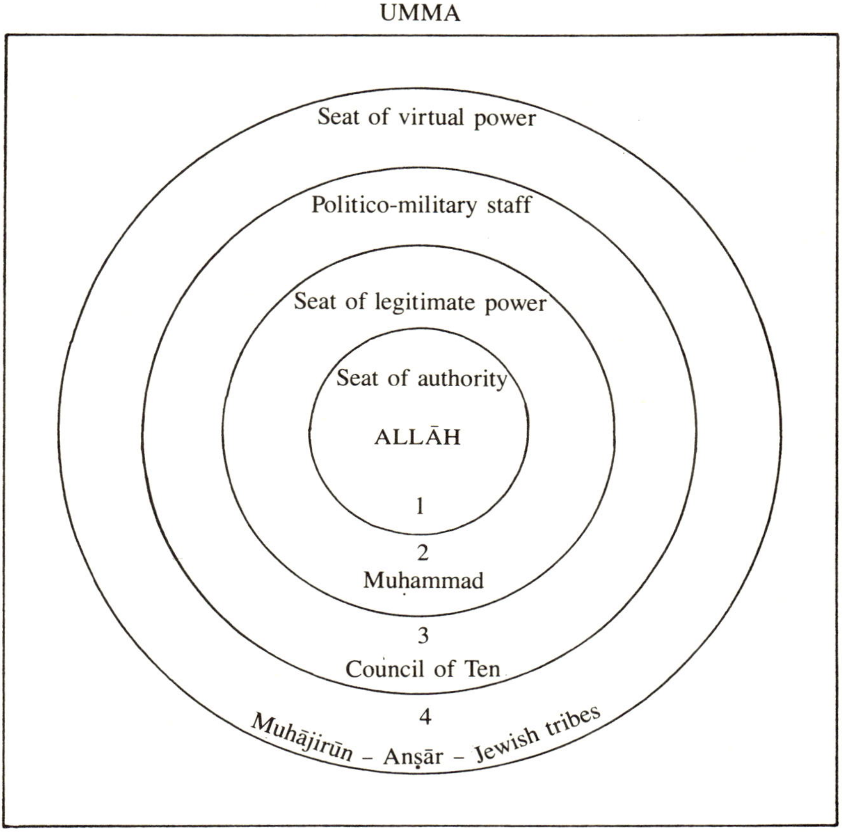

Figure 1. The Medina Model

that his works were at the apex of generalization, universalization, and mystification.

The Caliphate model: The Sunnī doctrine of power

The political principle of the caliphate (*khalāfat*) is simple enough: the leadership of the *Umma* must not be left vacant after the decease of the Prophet.[26] Its vacancy means that someone must be appointed to the supreme magistracy, the principal task of which is to regulate the affairs of the community. The first question is therefore that of the *obligatory and necessary nature of the caliphate*.

The necessity of the caliphate

By far the majority of the Sunnī theoreticians maintain the obligatory nature of the caliphate. Only a minority, of the Mu'tazila tendency, theoreticians such as Abū Bakr al Aṣamm and Hishām al Fuwātī, refuse to recognize its obligatoriness. For al Aṣamm, "the ideal community is a community of just men, which therefore has no need of political leaders. It is only the injustice of men that obliges them to give themselves, or to accept, a leader. Persons who refrain from destroying one another have no need of an imām."[27] Al Fuwātī in turn thinks that "when the Muslim community is disunited and plunged into anarchy, the entire imāmate (in its broad sense, including the caliphate) is an imāmate *de facto* and not *de jure*. It is only when the community, perfectly united, can reach an agreement on the choice of an imām, that this imām is legitimate."[28]

The theoreticians of the majority tendency who accept the necessity of a caliphate are divided into two groups. The first bases its necessity on scriptural evidence (*shar'ī*). The dogmatic schools, such as Ash'arī, take this position.

They base their belief on the Koran ("Oh you believers, obey Allāh, obey the prophet, and those among you who hold the authority"—IV, 62/59) and on tradition (*sunna*), for example, this *ḥadīth*, given by the Prophet: "Others after me will command you. A pious man will command you with his piety, the perverse man with his perversity. But listen to them and obey them in everything that conforms with the truth. If they do good, that will serve you and that will serve them. If they do evil, that will serve you and that will do harm unto them."

Ibn Taymiyya (d. 728/1328), the illustrious representative of Ḥanbalism, also holds that a leader of the Community is an absolute necessity, and in support of his position he quotes a *ḥadīth* from the Prophet, which says: "When three men depart on a journey, one of them must be their leader." Commenting on this, Ibn Taymiyya adds that "therefore the prophet ordained that there should be a leader at the head of the smallest group of voyagers, he wished to forewarn men of the duty incumbent upon them when they are in society."[29] This is the meaning of the sayings: "The sultan is the shadow of God on earth" and "Sixty years with an unjust imām are better than a single night without a sultan."[30]

Other theoreticians, more numerous, base the binding nature of leadership on a combination of the law (*shar'*, derived from *waḥy*—

revelation) and reason. This is the position of theoreticians such as Al-Māwardī, who says: "It is the nature of reasonable men to entrust (*taslīm*) themselves to a leader (*zaīm*) who will prevent them from committing injustices toward one another and who will adjudicate their quarrels. Without a leader, in effect, men would live in anarchy and indifference like savages abandoned to themselves."[31] Ibn Khaldūn expresses the same opinion, although he places more stress on the rationalist aspect of the necessity of a caliphate. "All society," he says, "must have a moderator (*wāzi'*) who governs it and who constitutes a source of appeal."[32]

The next question, the *appointment of the caliph*, has two aspects: the requisite conditions for candidacy to the caliphate and the forms of appointment themselves.

The requisite conditions

Muslim theoreticians have drawn up numerous conditions, similar to one another in their broad contours. A candidate to the caliphate must fulfill the conditions of maturity (puberty), reason, manliness, and freedom (thus slaves are excluded). Further, the most essential condition, on which both Sunnī and Shī'ī agree, is the *qurayshī* lineage. This fact is important since all four Arab caliphates so far—the caliphate of Rāshidūn (Orthodox), the Umayyad caliphate, the 'Abbāsid caliphate, and the Fāṭimid caliphate—all belonged to the tribe of Quraysh (i.e., the tribe from which Muḥammad himself had issued). The Ottoman caliphate is the one exception to this rule, for the simple reason that the Ottomans were of Turkish origin from Central Asia. Paradoxically, the Ottoman exception confirms another rule, namely that most dynasties claiming Islam as their authority have been *tribal* dynasties. There is thus a vast distance between the "universalist" discourse of the Koran and the historical reality likewise claiming a Koranic parentage.

In fact, the Koranic discourse has been used over the course of history by those holding power as an ideology to legitimate the absorption of non-Arab peoples and ethnic groups into their kingdoms.

Ibn Khaldūn, who was the first to perceive the discrepancy between tribal exclusivity and the universality of the Discourse, provided an admirable theoretical explanation for it, appending a sociological justification as well: "Royal authority," he says, "derives from superiority, which comes from the spirit of the clan (*'aṣabiyya*). *It is only the grace of God, in founding His religion, that unites individual hearts and vows to bring to bear their claims and aspirations.*'' He then adds that ''the religious structure eliminates competition and envy from among those united by blood ties, and turns all hearts toward truth.''[33] It bears mentioning that, while it is true that the machinery and reality of ''Islamic'' power lie in the control of the clan or the tribe, they were by no means its source of legitimacy for this power. Clan or tribal power derived its legitimacy from the Koran, which of course transcended clan or tribal divisions.

Forms of appointment

Islam in its original form provided no form or precise procedure for appointing the successor to the prophet. This in any case is the opinion of a large number of Muslims, the Sunnī.

The nonexistence of *naṣṣ* texts instructing on the choice of leader gave rise to a variety of forms and procedures, and the four Rāshidūn caliphs who reigned successfully over the community for twenty-nine years (632–661 A.D.) were, significantly, each appointed in a different way. If we add heredity (*irth*) to these four ways, there are five principal ways of appointing the caliph.

1. Deliberation in assembly (vestibule/saqifa). After the prophet's death, the Muslims had no clear idea how to proceed to appoint his successor. The constellation of forces gathered and the nature of the relations among them were such that the various groups (Anṣār, Muhājirūn, tribal chiefs, etc.) finally opted for the assembly procedure, which had in fact been traditionally practiced by Arabs before Islam. Congregating in the Vestibule of Banū Sā'ida, the most notable of Muslims (with the exception of the Hashimī and the ʿAbbāsids), with the aid of painstaking subterfuges and after numerous stormy discussions, finally reached a compromise. Later this compromise would be called *ijmāʿ* (*consensus omnium*) and would become one of the fundamental sources of Muslim law.

Abū Bakr was ''elected'' to the presidency of the community. It is of course pointless to compare this assembly procedure with that practiced in the modern democracies. The assembly that ''elected'' Abū Bakr was ultimately the work of the rather small groups who happened to be at Medina at that moment. It is known that the prophet's own family was absent from this assembly, and it goes without saying that non-Muslims were neither invited nor consulted. All things considered, however, the assembly procedure was probably the freest way of appointment, with the largest number of the Muslims present participating in proportion to the size of their respective groups. The other appointment procedures that would be used later in the history of Islam either would be arbitrary and hereditary (which amounts to the same thing), or would be the work of a very limited number of persons who had themselves been appointed by the presiding caliph.

2. *Nomination by the caliph in office ('ahd)*. This was how 'Umar, the second caliph, was designated. He was appointed by the unilateral act of Abū Bakr—the vow of allegiance (*bay'a*) would be a mere formality.

Muslim jurists of the Sunnī tendency have generally accepted the right of the caliph in office to designate his successor. For Al-Māwardī, "the act of nomination (*'ahd*) was one of the powers of the caliphate."[34] Ibn Khaldūn, sharing this view, specifies that "this right of succession (*walāyat al-'ahd)* was an integral part of the religious law and the consensus of the Muslim community specifies that it was permitted and accorded."[35] Ibn Khaldūn developed several arguments on this subject, and ascribes to the caliph in office an absolute right in this matter. "No suspicion," he says, "is justified in this matter, even if the imām (caliph) makes his father or his son his heir."[36]

3. *Nomination by a conclave (shūrā)*. The third designation procedure involved the establishment of a small conclave whose members were appointed by the caliph in office. This was the procedure whereby 'Uthmān, the third caliph, was appointed. 'Umar, stabbed to death by a Persian slave, appointed six persons from among the most notable before he expired.[37] Their task consisted in forming a council to deliberate the issue of succession. 'Umar had given the president of this council, 'Abd al-Raḥmān b. 'Awf, a decisive voice.

4. *The caliphate resulting from a revolt (fitna)*. 'Uthmān, the third caliph, was assassinated in the aftermath of an uprising. His successor, 'Alī, the last of the four Orthodox caliphs, was brought to power by the insurgents. This was the first time that a caliph had come to power on the basis of a decision made by the masses. But it would not be the last.[38] 'Alī was later himself assassinated by a Khārijī ('Abd al-Raḥmān b. Muljam Murādī). Thereafter, the designation of the caliph would be almost exclusively hereditary.

5. *Heredity (irth)*. After the extremely brief caliphate of Ḥasan, elder son of 'Alī, Mu'awīya b. Abū Sufyān of the Umayyad clan became caliph; with his accession the institution of the caliphate changed its nature, at least as regards the designation procedure.

Henceforth heredity (family lineage) would determine the person of the caliph, and in an analogous vein, monarchy would replace the caliphate of the first period of Islam. Ibn Khaldūn traces this transformation in detail. "The first Muslim," he says, "who sat on a throne was Mu'āwiya. He was distinguished by his corpulence . . . Later the 'Abbāsids, Fāṭimids, and all the Muslim sovereigns of east and west would have their thrones, their chairs and their seats, which would eclipse those of Khosro (Persian) and Caesar (Roman)."[39]

These five classical forms for the passing on of power were all subordinate to the principle of the *bay'a* (oath of allegiance), the Ottoman Empire again being the exception. According to Ibn Khaldūn:

> The oath of allegiance consists in paying the tribute of obedience. The person who gives his allegiance makes a kind of contract with his emir, upon whom he confers the government of his affairs and of all Muslims. He consents to recognize his authority and to carry out all his instructions, regardless of whether they should please him.[40]

But such a definition of the *bay'a* would seem rather subjective. In point of fact, the *bay'a* was rarely considered a *constitutive act*; ultimately what counted was the degree of force and the political power of the person who assumed the function of the caliphate. The *bay'a* was often merely a secondary consideration. The opinion of Ghazālī is illuminating in this respect:

> The choice of one man will suffice for the designation of the imām (caliph) if that man himself is an obeyed (*mutā'*) and powerful (*dhū showka*) person. If this result may be achieved through the allegiance (*bay'a*) of *one* single one among his electors, and if to arrive at this the agreement of *two* or *three* men who are equally *powerful* is lacking, efforts shall be made through negotiation to obtain their agreement on the designation of the imām (caliph). What counts is not the number of those who participate in this *bay'a*, but the legitimate power (*showka*) placed in the hands of the imām by virtue of this recognition.[41]

The imāmate model: The Shī'ī doctrine of power[42]

In the Shī'ī doctrine, the seat of power was occupied by the twelve imāms after the prophet. The principal difference between this doctrine and the doctrine of the Sunnī is that the imāmate may at any moment become a reality. It is potentially alive and present. In contrast, the caliphate in the Sunnī doctrine is an institution which has had its day. In other words, Sunnism can survive very well without the institution, and even without the idea of the caliphate, which would be absolutely unthinkable among the Shī'ī. For them, the idea, much more than the institution of the imāmate, is paramount since it is an integral part of imām doctrine. Without this idea, imāmism loses its *raison d'être*. This shows how important the idea of the imām and the imāmate is for the Shi'ism of the Twelve.

According to the Shī'ī faith, the Twelfth Imām (Mahdi), also called the Absent Imām, the Imām of Time, the Standing Imām, is still living and still the legitimate holder of power. He will reappear when God decides the appointed time. Since the imām is alive, the institution, or rather the idea, of the imāmate is alive, and so also is imāmism. Consequently it cannot be said that the imāmate model fits entirely into the traditional model of Islamic power. It would not be wide of the mark to say that the Shī'ī model is both traditional and modern at the same time, or alternatively, that it was a traditional model that has persisted into the present.

Thus Shi'ism attaches more importance than Sunnism to the necessity of a leader for the Community. So strong is this conviction that the Shī'ī persist in the belief that the prophet did not leave this world without having expressly designated his successor. This was 'Alī, son-in-law and cousin of the Prophet. Otherwise Shi'ism concurs with Sunnism in acknowledging the necessity of a government for the Community.

Necessity

The necessity of a political organization at the head of the Muslim Community is accepted by the Imāmiyya who, unlike the vast majority of Sunnī, believe that the necessity of an imāmate derives from religion itself (*wūjūb shar'ī*). Reason offers only a secondary justification of this necessity (*wūjūb 'aqlī*). To these two elements of justification the Shī'ī add a third, namely, the principle of Grace (*aṣl-é lutf*), whereby the imām is designated by the grace of God through the intermediary of his emissary. The Sunnī reject the idea of grace.

The requisite conditions

There are basically three specifically Shī'ī conditions.

1. Whereas membership in the tribe of Quraysh constitutes the condition *sine qua non* for accession to the caliphate in the traditionalist Sunnī doctrine, the Imāmiyya believe in *family lineage*, in particular, the family of the prophet (*Ahl al-Bayt*). The Prophet's family, of course, is descended by definition from the Quraysh tribe, and more particularly the Hashimī clan (*Banū Hashīm*). However, tribal membership is of secondary importance for the Shī'ī, the family being primary. The imām, therefore, must necessarily belong to the family of the Prophet. Further, the family line passes through the Prophet's daughter in the person of Fāṭima.[43]

2. In addition to having the correct family lineage the imām must also be infallible (*ma'ṣūm*). The notion of infallibility (*'iṣma*) occupied a very important place in the political and theological doctrine of Shi'ism. Infallibility is defined as "the faculty of being able to avoid sins or errors" (*Al-Munjid*). For the Imāmiyya, there are fourteen infallible beings: the Prophet, his daughter Fāṭima, and the Twelve Imāms. Once again, it was Divine Grace that bestowed infallibility upon the Quorum of the Fourteen. To bear out the point, the Imāmiyya refer to a Koranic verse that says (to the Prophet): "Have we not opened your breast and placed the burden weighing upon your shoulders far from you?" (CIV, 1–2).

3. Infallibility goes hand in hand with absolute moral integrity (*afḍaliyya*), a third specifically Shī'ī condition. This quality rests on two pillars: *taqarrūb* and *tafaḍḍul*. The first signifies that the imām must be the most worthy unto God, the second implies that the imām must be superior to others in respect of moral and political qualities.

The principle of absolute moral integrity has been utilized by the Shī'ī to vindicate 'Alī's prior right to succession after the Prophet's death.

The procedure for designating the imām

Unlike Sunnism, which has various ways to designate the caliph, Shi'ism has only one procedure for designating the imām. It is a hereditary form based on *predestination*. According to the Shī'ī faith, Allāh appointed 'Alī to the imāmate just after the creation of Adam. He was notified of the act of appointment later by Muḥammad. 'Alī in turn was to designate his elder son Ḥasan, and the latter his brother Ḥusayn for the imāmate. Then Ḥusayn designated his son and so it went up to the Twelfth Imām. The imāmate was thus passed from father to son (agnate lineage). But there were three exceptions to this rule: the Third Imām (brother to brother), the Seventh Imām (from the father to younger son, the elder son having died before the father), and finally the Twelfth Imām (also from father to younger son, although the elder son was alive).

The permanance of the imāmate

The extraordinary dynamism of Imāmism derives from two principal factors. First, Imāmism is based on a claim of a *political nature*, specifically the Prophet's immediate succession. This status has enabled it to assume the quality of a permanent challenge within the Muslim community. Second, this claim and its dynamism are continually refueled by the notion of *occultation*, which implies the permanence of the imāmate. Sunnism, on the other hand, which does not believe in occultation, remains on the defensive. Further, Sunnism has been too broadly identified with the institution of the caliphate throughout history, and has always had difficulty venturing beyond its traditionalist framework to renew itself and so be able to present a fresh political doctrine, more in accordance with changing circumstances of the times. In contrast to this rigidity, Shi'ism, which believes that the Absent Imām will one day reappear, has for all practical purposes been unencumbered by this sort of doctrinal problems. The imāmate continues, the system is reproduced, and above all faith in the permanence of the imāmate helps to spark the mobilization of the popular masses, by no means an insignificant consideration.

According to the Shī'ī faith, the Twelfth Imām has been hidden since the death of his father, an event that was supposed to have taken place in 871 A.D. Since then, he has lived a life in concealment. He will reappear when God so wills it. Two kinds of occulation (*ghayba*), minor and major, have occurred.

During the minor occulation (*Ghaybat al-Ṣughrā* 871–940), the Hidden Imām was represented by four specific delegates (*nawwāb khāṣṣ*) in succession. The disappearance of the fourth and last delegate marked the end of the cycle of the minor occultation, which then gave way to the major occultation (*Ghaybat al-Kubrā*). During the latter the imām had no special delegates, and the affairs of the Shī'ī Community were thus to be regulated by the General Delegates (*nawwāb 'āmm*). But who were they to be? This question gave rise to a lively theoretical discussion and even a bitter political struggle between two opposing

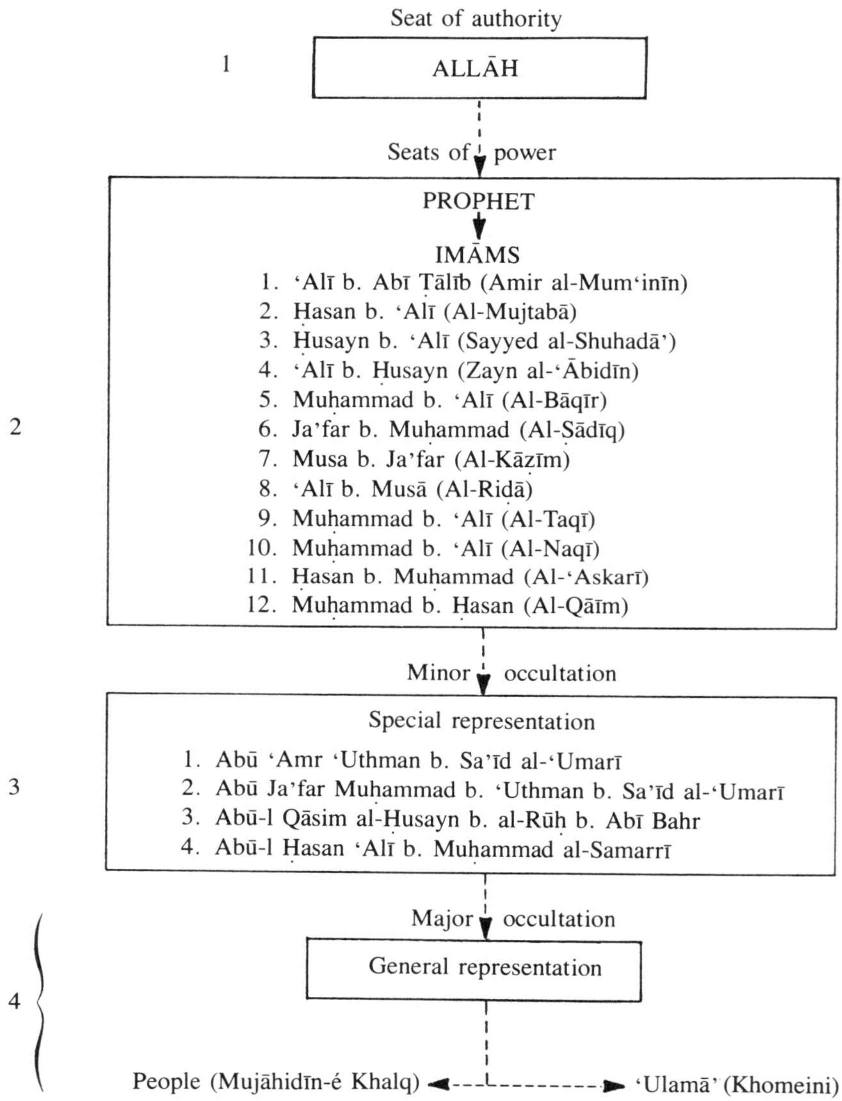

Figure 2. The Transmission of Shī‘ī Power

theses. The great majority of the *'Ulamā'* Shī'ī believe that the general delegation of the imām belongs by right to themselves because they are the most learned among the Shī'ī. They will surrender their power to the Hidden Imām as soon as he reappears. The idea is by no means merely fanciful. Ayatollah Khomeini reigns in Iran precisely in the name of the Hidden Imām, and in his quality of Delegate of the Imām (*Nāyeb-é Imām*).

The contrary thesis is maintained by those who wish to give a broader interpretation to the principle of delegation, and who accordingly hold that the act of delegation can be the monopoly of no one, and that it is the *Umma* or the *people* who are entitled to be the general delegate of the Imām. The Mujāhidīn-é Khalq in Iran hold this view, which undoubtedly accounts for the bitterness of their struggle to oppose Khomeini (to be discussed below).[44]

The Umma model: The Khārijī doctrine of power

The birth of Khārijism during the war of the Ṣiffīn (*ṣafar* 37/July 657) between 'Alī and Mu'āwiya explains in large measure why the Khārijī regard the *Umma* (Community) and not the caliphate or the imāmate as the seat of power. In that war, when Mu'āwiya's luck seemed to be wavering, he proposed to the opposing army a cessation of hostilities and recourse to arbitration to put an end to the conflict, a rather crafty maneuver, which put 'Alī in a predicament and provoked dissension within his army. 'Alī, caliph of the Muslims, could not easily refuse arbitration (*aṣl-é taḥkīm*), which in fact is recommended by the Koran. He therefore resigned himself to accepting the enemy proposal. His decision promptly caused part of his army to protest. Reciting one of the Koran verses (*Lā ḥukma illa Lillāh*—No decision except with God),[45] the dissidents thereupon left 'Alī's army. This was one reason they received the cognomen *Khārijī* (the Leavers).[46]

The Khārijī went on to fashion a political doctrine that distinguished itself by its refusal to recognize the Qurayshī claim to the monopoly of power. Membership in the Quraysh tribe had then as well as later (until the end of the 'Abbāsid empire) been one of the necessary conditions for access to the caliphate. It should, moreover, be borne in mind that the Ṣiffīn war was in a certain sense a war between two rival Quraysh clans: the Hashimī in the person of 'Alī, and the Umayyads, represented by Mu'āwiya.

At this point some Muslim soldiers decided to summarily dismiss the two illustrious representatives of the two opposing clans. This proved a major event. Not only was tribal membership challenged, but also Arab lineage (or Arab ethnicity) was also taken under advisement by at least one of the innumerable Khārijī fractions (the Azraqiyya).

In refusing to recognize Qurayshī lineage as a necessary condition for access to the caliphate, the Khārijī had taken, perhaps even unconsciously, a major step toward the universalization of Islamic power. According to the Khārijī thesis, the *Umma* alone, collectively, is entitled to choose its head, beyond any tribal, familial, or other contingencies that may arise.[47] This idea, which was an attempt to restore the *Umma*'s status and place it at the head of the hierarchy of earthly authorities in the post-Prophet period, and went so far as to situate it squarely in the legitimate seat, the very source of legitimation, remarkably enough found no favorable resonance for centuries. The idea did not appear again among Muslim thinkers until the nineteenth and twentieth centuries. Interestingly, this time the people and the movements that would take up this originally Khārijī idea were intent on camouflaging the primary source of their inspiration. It is, in fact, their adversaries who accuse them of being "shameful Khārijī" (e.g., Khomeini's accusations against the Mujāhidīn-é Khalq and their theoretician, 'Alī Shar'īatī, which will be discussed below). This will perhaps permit a better understanding of the climate of terror in which Khārijism, and ideas like it aimed at freeing the *Umma* from ethnic, tribal or family constraints, have had to subsist.

The modern Khārijī also have another source of inspiration outside of Islam, namely, in the politics and culture of the West. Concepts such as *nation* and *people* in which the Western democracies are rooted, are not *a priori* so very far from the Islamic concept of the *Umma*, and these modern Khārijī have therefore attempted conceptually to reconcile the *Umma* with the People, and the Community with the Nation, with a view toward establishing an organic link between divine sovereignty, on the one hand, and national or popular sovereignty on the other. It may be argued that therein lies purport of the challenge that certain "Muslim modernists" have addressed to their "traditionalist" coreligionists.

The seat of virtual power

This sphere of virtual power encompasses the totality of emirs, princes, sultans, generals, and generally speaking all those who have reigned *in their own name* over some part of the Muslim empire. These wielders of virtual power, who were not appointed beforehand to the "posts" they occupied, owed their position to their own strength. Furthermore, this group was totally distinct from the caliphate establishment. Of course, a number of them may on occasion have served the caliph, either personally or through their families. This was the case of the Tāhirid, the Ghaznavid, and the Seljuqid. Nonetheless, as soon as they established themselves as a legitimate power *in their own right*, they lost no time in breaking their ties with the administrative establishment of the caliphate and establishing their own *de facto* autonomy in its place.

Māwardī, whose life spanned the period of this transformation of local administrators into a *de facto* autonomous power—an event that had become commonplace because of the structural weakness of the caliphate—drew a clear distinction between these two types of situation. There were, according to him, two types of *emirate*: emirate by designation (*imārat al-istikfā'*) and emirate by usurpation and domination (*imārat al-istilā'*).

An emirate by designation, in which the pact was concluded with a person freely chosen by the caliph, entailed circumscribed territorial powers and material powers by covenant. An emirate of this type was conferred when the caliph assigned the emir authority over a city or a province (*wilāya*), investing him with powers to be exercised in specified administrative matters. In these matters, and within these limits, the emir had full authority. There were seven areas under his authority: the organization of the armed forces, administration of justice, the collecting of taxes, defense of the faith, application of penal measures, conducting prayers, and the organization of pilgrims.

An emirate by usurpation was established when an individual had seized certain provinces by force, and the caliph conferred upon him the emirate over these provinces with the mandate to administer and govern.

In this case, the emir, by the very fact of *seizure*, enjoyed full powers in matters of administration and government, and the caliph, by the fact of investiture, endowed his power with a legal quality, thereby validating the irregular situation and transforming a state of interdiction into the permissible. Most certainly, this type of investiture departs from the rules governing investiture by common law as regards both the conditions of entitlement and the scope of the investiture; but it ensured the safeguarding of juridical and religious principles, which were to be neither transgressed, violated, corrupted, nor weakened. Thus, in investiture by usurpation and constraint *some things* volens nolens *were deemed permissible which would not be tolerated in an investiture by designation and free choice*—the difference, namely, between power and weakness.

For the usurper, the legal consequences of investiture are seven in number:

—The first is to effect his recognition of the authority of the caliph as the vicar of the prophet (seat of power) and spiritual sovereign;

—The second is to reduce the usurper to obedience, ending the condition of insubordination and expunging the sin of revolt;

—The third is to establish an understanding with a view to friendship and assistance (among Muslims);

—The fourth is to give permitted status to the conferments of such emirs, and executive status to decisions made by those occupying these emirates;

—The fifth is to ensure the regular collection of taxes;

—The sixth is to ensure a regular application of penal provisions;

—The seventh is to establish the emir as defender of the faith.

After having laid down these rules, Māwardī concludes that "When the usurper meets all the conditions of competence required by investiture on his own free will, there can be no avoiding granting him the investiture so as to bring him to obedience and put an end to schism and revolt."[48]

Its legality thus established, an emirate by usurpation differs in four respects from an emirate assumed by designation:

1. An emirate by usurpation naturally falls to the person who accomplished the act of conquest, while an emirate by designation is validly acquired only by the person chosen by the caliph's will;

2. An emirate by usurpation gives powers both ordinary and extraordinary, while an emirate by designation is invested only with ordinary powers;

3. An emirate by usurpation admits of a vizierate by delegation, but an emirate by designation does not.[49]

Finally, we may recall the role played by the *'Ulamā'* and the *Fuqahā'* (intermediate instance) in the process of legitimation and legalization of an emirate by usurpation. The *'Ulamā'*, a privileged group, beneficiary of the "Islamic" quality of the empire, generally tended to facilitate the integration of the wielders of virtual power into the caliph system of which they were the defenders. This was done by swearing an oath of allegiance to the emir in the form of a *fatwā* or an act, i.e. legitimate collaboration with the emir, on the principle that *silence implies consent*. The latter case, which was extreme and rare, applied to a situation in which the *'Ulamā'* were unable openly to grant the emir their guarantee because of his blatant violation of *sharī'a*.

The intermediate seat

The intermediate sphere is covered by the *'Ulamā'*.[50] Historically, we still do not know exactly when and how a group of Muslims were able to constitute themselves into a distinct body and assume the role of intermediary between the *sharī'a* and Muslims. We know that in the first period of post-Muḥammad Islam, or more precisely at the time of the four Orthodox caliphs, it was the Companions (*Ṣaḥāba*) of the Prophet who fulfilled the function of interpreting the revealed texts and the tradition of the Prophet (*sunna*), in addition to their political, military, or administrative duties. In other words, it was they who established bonds and they who severed them (*Ahl al-hall-é wal-'aqd*). As previously mentioned, the transmission of power and the procedure for legitimation of caliphal power were also among the "attributions" of the *Ṣaḥāba*.[51]

The problem changed complexion when the generation that had been the direct witness of the progressive evolution of Islam passed on. The next generation, called the disciples of the companions (*tābi'īn*), who had known the times and the works of the *Ṣaḥāba* intimately, undoubtedly set themselves up as the "authorized" interpreters and transmitters of the Koran and the *sunna*.

Thus constituted, the *'Ulamā'* justified their mission by the following verse from the Koran: "It is not for the believers to march out all together; so why should not a party from each section of them march out, in order that they may gain understanding in religion, and that they may warn their people when they return to them? Perhaps they will beware?" (IX, 123/122).

Whatever the legal underpinnings of the *'Ulamā'*, it is their historical reality that is important in the present context. And this reality attests to the existence of a specific group that performed (and still performs today) important functions within the community. The interest here is in their political role, which is essentially the role of intermediary within the sphere of legitimate power, as well as between legitimate and virtual power.

Thus, the domain in which the *'Ulamā'* exercised their authority was, depending on the particular case, the *caliphate*, the *sultanate*, or the *emirate*. Here one more distinction arises between the Sunnī model and Shī'ī model. In the Sunnī model, the *'Ulamā'* played a relatively more important role than their Shī'ī colleagues, at least down to the fall of the 'Abbāsid empire (1258 A.D.), in particular as legitimizer of whomever was wielding power, or exercising virtual power.

In reality, although he was Qurayshī and appointed by his predecessor, the caliph was obliged by circumstances to give his power a purely religious quality. The reason was twofold: first, to surmount the tribal and familial divisions and circumstances had imposed upon him (the Khaldūn theory), and second to consolidate his hold on power and thereby expand his empire. The "cause of Islam" was a useful argument in both cases.

The *'Ulamā'* were the sole interpreters of the *sharī'a* and hence the only group able to accomplish this mission. Muslim historiographers have indeed provided us with numerous illustrations of how they did so. Let us take, for example, the ceremony of the investiture of the 'Abbāsid caliph Al-Muqtadī (467–487/1075–1094). Ibn Kathīr reports that

> the traditional ceremony of the *bay'a* took place at Baghdad in *sha'bān* 467/April 1075. The first to swear an oath of fidelity to the caliph was Ṣerif Abu Ja'far, whose affection assured Al-Muqtadī the support of both militant Ḥanbalism and of the great masters from the population (*'amma*) of Baghdad. Then came the turn of the two most eminent *shafi'ī* doctors of Baghdad, Abū Ishaq al-Shirāzi (d. 476/1083), the director of the Nizāmiyya, and his rival Abū Nasṛ b. al-Ṣabbāgh (d. 477/1084).[52]

This example shows clearly that the Ṣerif's political support of the caliph was not enough in itself and that the caliph could ill do without the guarantee of the *'Ulamā'*, expressed solemnly and publicly. With that guarantee, the caliph could be sure that the way was barred to any challenges of a juridical nature to this authority.

Of course, the degree and the weight of the influence of the *'Ulamā'* depended on the personality and political power of the caliph. For a caliph who was strong politically, the justificatory act of the *'Ulamā'*, in the form of the *bay'a* or the *fatwā* (issuance of a writ of juridical and political relevance on the *sharī'a*) had more of a *declarative* and *symbolic* value. Conversely, the same act could also have a *constitutive* and *real* value if the caliph was not sufficiently strong.

As regards the sultanate and emirate, the attitude of the Sunnī *'Ulamā'* was generally realistic. With the 'Abbāsid empire disintegrating, they accepted the existence of sultans and emirs as a fact. While taking the precaution to confirm the authority of the caliph in power—at least at the formal level—the *'Ulamā'* endeavored to bring the *facts* in line with the *law*. As described by Ghazāli:

> The caliphate belonged to the Banūl-'Abbās and to attempt to take it from them was out of the question. *De facto* power in the various provinces belonged to the sultans or kings. To challenge them was equally out of the question; to attempt to overthrow them would lead to serious disorders; furthermore, the Prophet had commanded obedience to emirs who held power. The solution lay in a compromise. Royalty would acquire legitimacy for itself by recognizing the preeminence of the caliph's authority. By coming to terms with royalty, the caliph acquired—at least in part—the *de facto* power which he would have lacked without this support.[53]

Through compromises of this sort, the institutions of the empire were able to maintain relative stability and harmony.[54] Once the sultan or emir was legitimated (by the *'Ulamā'*) the caliph had only to bestow upon them honorary titles, often having a religious connotation (for example *Rukn al-Dīn*—Pillar of Religion, in addition to the title *Shahinshah*, attributed to Malek Shah Seljuk [d. November 15, 1092]). Let us be fair to this general political strategem of the *'Ulamā'*: they were in this way able, despite the turmoil, to maintain the balance of power between the caliph and the various princes; furthermore, the latter were often from a different ethnic group, spoke a different language, and in some cases were even of a different religion than the caliph in power. Thus, in the final analysis it may be fairly said that the *'Ulamā'* had succeeded in their task, in many respects a difficult and delicate one. It is furthermore an established fact that the fall of the 'Abbāsid empire was not due to the internal disintegration of its institutions: the blow came from without, from Central Asia, to be exact.

The Shī'ī *'Ulamā'* adopted a different approach, and the difference was the natural consequence of the fundamental difference between Shī'ī and Sunnī theories of power and politics. In Shī'ī theory the person of the imām is considered infallible. With the imām present, the *'Ulamā'* have no practical or theoretical role to play in the legitimation or transmission of power. Under the imām they can only be his partisans or disciples. In a sense they are the companions or scribes of the imām. The situation can change radically if the imām is absent, which can take place only once. During his absence, the *'Ulamā'* bring their full political weight to bear, in this going much further than their Sunnī counterparts.

During this period, which in principle can last forever, the Shī'ī *'Ulamā'* have made use of two strategies. The more moderate have chosen a strategy quite similar to that adopted by the Sunnī *'Ulamā'*, namely, the role of intermediary between the seats of legitimate power and virtual power. Thus, they certify, justify, and confirm the legitimacy of princes having the reality of power in their hands. These princes may be Shī'ī or claim to be Shī'ī, or in certain cases might not be Shī'ī. Shī'ī *'Ulamā'* have even on occasion served non-Muslim princes. This was the case with Naṣīr Tūsī, an eminent Imāmiyya theoretician and author of a number of theological works; this did not prevent him from personally accompanying the Mongol Khan Hūlakū to Baghdad and becoming his adviser. Hūlakū marked the end of the 'Abbāsid empire.

The other strategy, more radical, consisted in entering into direct conflict with the prince to deprive him of power. Although used less frequently, this strategy has been by far the most dynamic in history. It was through this strategy that Ayatollah Khomeini succeeded in overthrowing the Shah's regime.

3. Contemporary Models

Monist models

Models of a religious nature: The Islamic Republic of Iran

The question of the seat of authority is clearly and distinctly resolved in the 1979 Constitution. Sovereignty belongs *exclusively* to *the one and indivisible God* (second principle, §1). This principle is reinvoked and confirmed repeatedly throughout the text of the constitution.

Thus, theoretically, the problem is settled. The seat of authority is *one and undivided* and the intermediate seat (*'Ulamā'*) rules in the name of the seat of power (imāmate). The various seats are reduced to two, with the intermediate seat and that of virtual power disappearing. This original formula contrasts in certain respects to that legitimated earlier in the 1906 Constitution.

The 1906 model[55]

The seat of authority is not clearly defined in the 1906 Constitution; it does not stipulate where or with whom sovereignty resides. Indeed, the term for sovereignty itself (*ḥākimiyyat*) is not even mentioned in the Constitution. On the other hand, it does deal with the question of royalty (*salṭanat*), if in vague terms. "Royalty," says the Constitution, "is a divine trust (or Grace) bestowed by the Nation on the person of the Shah (Emperor)" (Article XXV of the Annex to the Constitution).

In rereading this article, one can appreciate the artful ingenuity of the writers of the Constitution. Their task was presumably to reconcile three principles that are not necessarily reconcilable, namely:

—the *salṭanat* royalty, a traditional Iranian principle, this time in the form of a trust;

—*divine trust* (*mūhibat ilāhi*), a providential principle in Islamic form (*Ilāh*);

—the Nation (*Mellat*), an autonomous and indivisible entity, a civil notion divested of any religious, ethnic, linguistic, or other connotation.

There is a fundamental difference between the principle of royalty and the other two principles.

Specifically, royalty is referred to in the Constitution as a trust conferred by the grace of God. Hence, it need be neither unchanging nor lasting. God and the Nation, on the other hand, are quite different.

Since the *trust* is not of divine origin, but conferred by the Nation on the person of the emperor, one may fairly ask: Does the emperor derive his powers from the *divine source* or from the *Nation*? The Constitution is silent on this point and, indeed, not unintentionally: it is a silence imposed by circumstances. At that time a bitter struggle was taking place between the *traditionalists*, led by the *'Ulamā'*, and the *modernists*, supported by the laity. The vague and ambiguous formulation in the above article represents a compromise between these two tendencies.

A reading of the 1906 Constitution confirms this. This Constitution is clearly *dualist*: Sovereignty resides in both God and Nation simultaneously but separately (Figure 3). In this diagram, God is *Allāh* and none other (the God of Islam) and is represented by his Religion, here in its imāmite form. The *'Ulamā'* are charged with guaranteeing the just application of religion. To do this they are empowered by the constitution to reject laws that might be in conflict with Dogma (Article II of the Annex).

The *Nation* is represented by its parliamentary delegates (in the Senate in the National Assembly). The 1906 Constitution gave the Nation a liberal framework (freedom of trade, freedom of profession, protection of private property, etc.).

It is clear that the Pahlavis wished to eliminate instance 1, and to achieve this they placed their reliance on instance 2 by strengthening b. Their great mistake was that they joined battle on both fronts: They attempted to weaken both genuine parliamentarism and clericalism at the same time. They failed in both.

The last Shah attempted to replace the nation as a sociological,

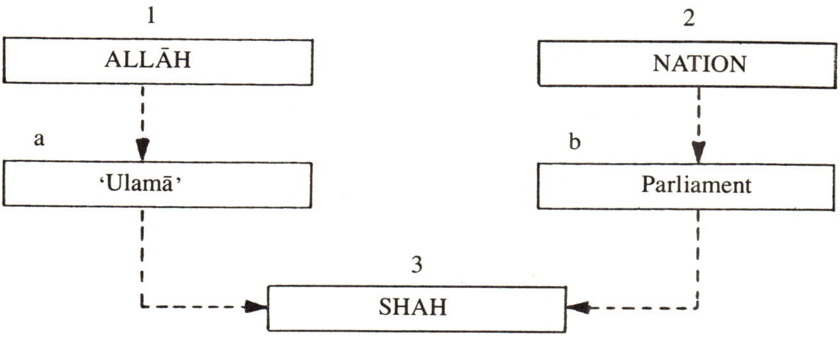

Figure 3. "Bilegitimist" Model of the 1906 Constitution

concrete entity with a historical and legendary Nation. He even actually ceased speaking to the living, and rather preferred to address Cyrius and Darius. Having chosen to maintain a distance between himself and the most active and influential force in the nation, namely the liberal (or national) bourgeoisie, especially after the fall of Mosaddeq in 1953, he remained socially and juridically isolated without any foundation of legitimacy to the very end of his reign.

By installing a *de facto* despotism and demolishing every basis of genuine parliamentarism, the Shah in the end played into the hands of the clergy. Indeed, he bestowed privileges on his most embittered enemies in the person of the extremist *'Ulamā'*, while shying from his natural allies. Thus, instance 3 was eliminated by a temporary collusion between the two instances 1 and 2. Later on, instance 2 was in turn eliminated by instance 1.

The 1979 model

> The Islamic government is subject to the Law of Islam, which comes from *neither the people nor its representatives, but directly from God* and his divine will.
>
> Ruhollah Khomeini

The constitution of the regime of Ayatollah Khomeini is a model of power in simple and clear form, unitary rather than dualist. Allāh is the supreme and unique seat of authority. "The Islamic government," says Khomeini, "is a government of Law. Sovereignty is limited to God, and the law is God's commandment. Everyone, the Prophet, his successors, and others, are subject to the law for all eternity. If the Prophet assumed the responsibility of representing God, it was only by order of the Lord.[56] Consequently, just as power was transmitted from Allāh to the Prophet, the Prophet must in turn transmit it on God's command to the imāms. The Twelfth Imām, who according to the Shī'ī faith is still alive, but absent, is the legitimate trustee of this power. In his absence, the representative (delegate) of the imām, Nāyeb-é Imām, is empowered to carry out the tasks of the leadership of the government. Indeed, Khomeini's own words say it to us clearly:

> The just imāms and Jurisconsults (*fuqahā'*) are under the obligation to utilize the organization and the formation of the government to apply divine prescriptions, establish the equitable regime of Islam, and serve the people. To them the bare fact of wielding power means but trouble and bother. But what else can they do? Their mission is to accomplish this duty, which is the very object of the tutorate of the Jurisconsults (*walāyat-é faqīh*).[57]

This statement is reproduced in its entirety once again in the 1979 Constitution in its fifth principle.

The fifth principle stipulates that the delegate of the imām must be "accepted as leader by the majority of the people." Theoretically, this formulation could also be construed to mean that a popular vote in the spirit of Rousseauian democracy would be decisive in electing the leader. In that case, the Khomeini model would fit the category of dualist models based on two instances of authority: God and the people, for example. But this is not the case, for three main reasons:

(1) This thesis presumes that sovereignty belongs to the people or the nation. But this is not the case in the 1979 Constitution. As stipulated in the Constitution (second principle), the only sovereign is the One and indivisible God, uniquely, absolutely, and definitely.

(2) The people cannot choose whom they wish. At best, they can elect their leader from among a restricted and specified number of *'Ulamā'*. Moreover, the candidates to the worthy title of leader are not candidates by virtue of their capacity as citizens but because they claim to be delegates of the Absent Imām. Clearly this vitiates any presumed democratic nature of the popular vote.

(3) Finally, popular and national sovereignty does not accord with the Shī'ī notion that imāms are delegated by Allāh, and thus are infallible (*ma'ṣūm*).

A careful reading of both the text of the Constitution and the writings

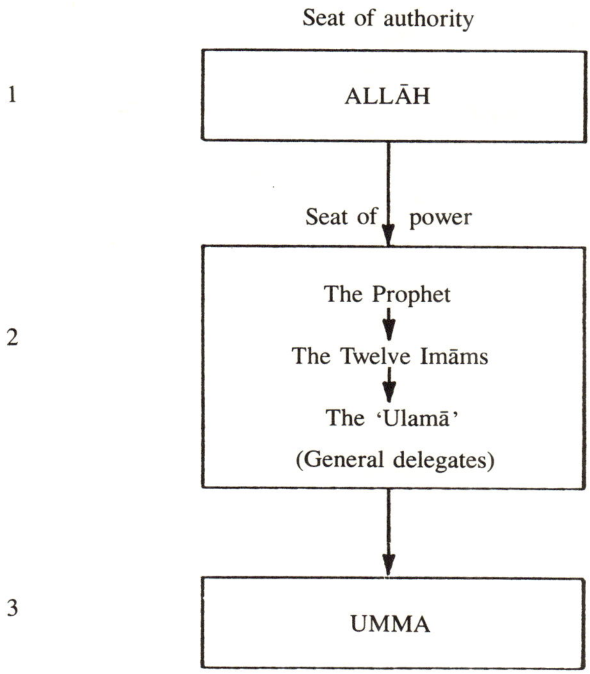

Figure 4. Khomeinist Model of Power

and declarations of Ayatollah Khomeini[58] are unequivocal in the rendering we have just given (Figure 4).

In this unitary model, real power is in the hands of the leader. He is above the legislative,[59] executive, and judicial "bodies" (a more fitting term for them than "powers"). Indeed, each of these bodies depends directly or indirectly on the leader, and acts under his orders. The vote of the people or "their" representatives has, it should be stressed clearly, a *declarative* and not a *constitutive* value.

To sum it up, what is original about the Khomeinist model is that the intermediary seat (the *'Ulamā'*) seized power to install a unitary model of its own preference.

Models national in essence: The Turkish Republic

The Turkish Republic is a unitary European-type model based on the nation state. In this regard, the Turkish regime would not *a priori* have been distinguished by any essential singularities if the model it had adopted had not had to function in a society in which the vast majority of the population (98%)[60] is Muslim, and which, moreover, had been the heart of the Ottoman empire and the seat of the last Muslim caliphate just before the secular model was installed. After the establishment of the Turkish Republic under Mustafa Kemal (born Thessaloniki, 1881, died Istanbul, 1938) this situation gave rise to a discordance between the *apparatus of power*, animated by a firm resolve to secularize the society, and the *social basis*, enveloped in a deep religiosity. A number of the sociopolitical conflicts and tensions that Kemalist Turkey would experience derived effectively from this discrepancy between the *political* and the *social*.

Since we are here concerned with an analysis of the seat of power, we shall attempt to place in relief the two factors, *national* and *secular*, which we think are the most distinctive features of the Turkish regime.

Turkish national sovereignty

The Ottoman Empire that preceded the Turkish Republic was a multiethnic and multilingual empire. Islam was the sole unifying force holding together the diverse peoples scattered over the immense territory of the empire and subject to the authority of the Ottoman Sultan-caliph,[61] whose person represented the continuity of orthodox Islamic power, at least in the eyes of the subject Muslim population.

The Turkish element was, to be sure, predominant in the Empire, as the Arab element had once been in the Umayyad and ʻAbbāsid empires. But the ethnic preeminence of the Turks was not in itself of a nature to foster nationalist sentiments among the Turks. Quite the contrary. With the reins of power in their hands, the Turks of course had every interest in making this vast and great empire as everlasting as Allāh permitted. The first task, therefore, was to oppose and contain by all possible means any separatist movement of ethnic, religious, or national character. And they were in the main successful in doing so, until the Ottoman Empire was shaken up irrevocably by World War I and its consequences. Thus as long as their empire endured, the Turks would be the last to resort to nationalism.

Since the nineteenth century certain Turkish elements had begun to challenge the authority of the Sultan-caliph, and the movement known misleadingly as the "Young Turks" played a prominent role in this regard. The "Young Turks"—who would have been better named

"Young Ottomans" inasmuch as their organization was called "The Ottoman Assembly for Unity and Progress" (*Osmanlı Ittihad ve Terakki Cemiyeti*)—aspired by definition to *Osmanism*, not to Turkish nationalism. Their goal was to modernize the institutions of the Empire to make it more viable. Founded in 1889 by students at the Military School at Istanbul, the Assembly likewise wanted to fortify—as in fact their name indicates—the *unity* (of the Empire), not to contribute to its decline and disintegration. There can be no doubt, therefore, that the "Young Ottomans" were *unionists* who aspired to take the fate of the Empire in their hands and to reign in the name of Islam. Indeed, when the Enver–Jemal–Ṭal'at triumvirate, progeny of the Assembly, came to power in 1912, the unionists did not hesitate to play the card of Islam for all it was worth. They even proclaimed a *jihad* (holy war) against the Russians, the British, and the French.

There was no very clear sense of nation to be found among the Ottoman Turks themselves on the eve of World War I. The National Pact (*Millī Misak*), which was adopted at war's end and served as a kind of manifesto of the nationalists, does not even mention Turkey as such, but speaks merely of where "the majority of Ottoman Muslims live." References to Islam, by contrast, are frequent; it speaks for example of "Muslim territories and populations" or the "Islamic caliphate and the Ottoman sultanate."

In fact, it was only after events themselves had intruded (most notably the Treaty of Sèvres on August 10, 1920) that the Turks, faced with the realities of the epoch (the territorial disintegration of the Empire) began to realize their international isolation, and that the only recourse left open to them was to unite under a national banner. Mustafa Kemal (who took the name Atatürk ["Father of the Turks"] only in 1934, when he passed a law ordering all Turks to adopt a surname) resolutely assumed the leadership of this movement and went on to build the young Turkish Republic on the ruins of the Ottoman Empire.

But the Republic was not proclaimed until October 24, 1923, for first Kemal had to effect a separation between the sultanate (royalty) and the caliphate (vicariate of the Prophet), and then abolish them in turn, beginning with the sultanate in November 1922. Henceforth, the Ottoman caliph was no longer regarded as a sultan (king) but merely as the repository of Islamic legitimacy. For sixteen months, from November 1922 to March 1924 (the date the caliphate was abolished), the Ottoman caliphate would thus find itself in the same situation as the 'Abbāsid caliphate had been in its last days. Real power was no longer in the hands of the caliph, but in the control of a strong man in the person of Mustafa Kemal, who would be elected President of the Republic on April 30, 1924.[62] With the transformation of the sultanate into the Presidency of the Republic, following the abolition of the caliphate pure and simple, the way was cleared to the formation of a truly national Turkish state.

These events were without precedent in the history of Islam. It was the first time that the dominant ethnic group, with imperial power and caliphal dignity in its hands, undertook on its own initiative to destroy the institution of the caliphate and replace it with a national and secular state. True, the Umayyad caliphate (of the Quraysh tribe) had likewise been abolished by other Quraysh (of the ʻAbbāsid clan), but the rise of the ʻAbbāsids could not have occurred without the invaluable assistance of other (non-Arab) ethnic groups, notably, the Khorasanids under the command of Abū Muslim. Furthermore, the ʻAbbāsids did not abolish the caliphate as an institution, but merely put an end to the reign of the Umayyads (in 756 A.D.) to establish another caliphate, which by definition was neither national nor secular. The ʻAbbāsid caliphate in turn fell under the onslaught of the Mongol hordes (in 1258 A.D.).

Turkish nationalism had other original qualities as well. It was the only nationalist movement that came to power and established a national state in a Muslim country without the aid of religious forces. In all other Muslim countries, national liberation movements have felt constrained to seek such an alliance, and the establishment of an independent state, where that has occurred, has been the fruit of this joint effort. But the consequence over the longer term has been ambiguity and confusion in state institutions, and serious political conflicts between the nationalists and the religious forces. The reason for the powerlessness of the national movement to disengage itself cleanly from the Islamic movement is no doubt to be sought in this "original sin" of establishing a common state with it.

The originality of the Turkish regime consisted in its success in doing just this.[63] Hence the religious forces found themselves looking on from the sidelines as the Turkish Republic was being molded and ultimately established, in many respects behind their backs and against their manifold interests. In political terms, therefore, modern Turkey owes neither its existence nor its evolution to religious forces, but is entirely the work of its nationalists.

The Republic proclaimed in 1923 became a national state in 1928.[64] Later Turkey, following the example of the rest of Western Europe, adopted the principle of *national sovereignty*. Indeed, its adherence to this principle has been so firm that neither the innumerable difficulties that have beset the regime, especially after the death of Atatürk in 1938, nor the two military *coups d'état* (in 1960 and 1980) have diluted its national character in any respect. In fact the *coups d'état* were effected in the name of the Turkish nation and to "reestablish the values introduced by Atatürk."[65]

Further, the national character of the Turkish regime has been clearly and explicitly spelled out in the four constitutions Turkey has had up to the present. The first (provisional) constitution, of January 20, 1921, which had only twenty-three articles in all, established the principle of *popular sovereignty* (Article 1). The other three constitutions would abandon the notion of popular sovereignty in favor of *national sovereignty*, and the next constitution (dated April 20, 1924 and amended seven times) accordingly defines the character of the regime as follows: "The Turkish state is republican, nationalist, statist, secular, and reformist" (Article 2). The second constitution was inspired by the Polish constitution of 1920, in the era of Jozef Pilsudski. The third constitution (January 1961) was in turn influenced by the constitution of the Federal Republic of Germany. While maintaining the national character of the state, the 1961 constitution gave it a social coloring with terms such as *Sosyal Hukuk Devleti*, literally "the legal social state." The 1961 constitution specifies that "The Turkish Republic is a nationalist democratic secular and social state . . ." (Article 2). This formula would be restated in similar terms by the writers of the most recent constitution. The constitution of November 22, 1982 specifies that the "Republic of Turkey is democratic and secular and social . . ." (Article 2), and Article 6 is careful to state that "sovereignty belongs to the nation unconditionally and without qualification."

In his oath of office, the Turkish President, unlike his counterparts in other Muslim countries, swears not on the Koran but on his own "honor and integrity," and not before God, but before the Great National Assembly (*Büyük Millet Meclisi*). He swears to undertake to "safeguard the indivisibility of the territory and the nation" and scrupulously to respect "the absolute sovereignty of the nation" (Article 103). It is obvious in such a system that neither is Islam the religion of the state, nor is the President of the Republic obliged to profess it, inasmuch as he is regarded as the guardian of the secularity of the state. Let us now examine this secularity.

The secularity of the Turkish regime

Like Turkish nationalism, which was the political response of ethnic Turks to the collapse of the Ottoman Empire in the aftermath of World War I, secularity was a reaction, in another form, to Islam as a system.

It seems evident enough that the secularization of a regime, which for centuries had embodied traditional Islamic power in the institution of the caliphate and furthermore was a formidable political and military power in its own right, could not have taken place without the tenacity and boldness of a man the stature of Atatürk, and without recourse to force and repression. But although the Kemalites succeeded in secularizing the power apparatus, they were not able to do as much for Turkish society, which has remained deeply religious in spite of all.

Secularization took place on two levels: in institutions and in symbols.

The secularization of institutions. At this level, the aim was to effect a total break with the fundamentally Islamic character of the Ottoman caliphate. This break was indispensable to the "de-Islamization" of the power apparatus. The sultanate and caliphate were therefore abolished successively, as mentioned. But bold as they were, these measures were not in themselves sufficient to transform Turkey into a totally secular state. Other measures were necessary, and the total revision of the 1924 constitution was one of these.

After the constitution was "purged" of its Islamic content, it was the turn of public institutions. First, the Ministry of Religious Affairs and Pious Foundations was abolished and replaced by simple "administrations." Religious instruction was also placed under strict controls, and after 1926 the Institute of Islamic Studies at the Faculty of Arts of Istanbul was the only school authorized to provide such instruction.

In a further step, Mustafa Kemal gave his country a civil code modelled on the Swiss code. Polygamy was prohibited, civil marriage was made mandatory, and inheritance laws were radically transformed.[66] The work continued, and numerous other measures ensued to "civilize" the institutions of the Republic.

The secularization of symbols. Every state has its symbols. The new secular state had to create its own symbols after having abolished those of the fallen caliphate. Indeed, it would have entailed grave risks for the young republic if it had permitted itself to indulge in the free use of religious symbols or symbols that were even suggestive of religion. Use of signs and symbols belonging to the Islamic past might also have fomented political hostility to a regime that had so loudly proclaimed its secularity.

The Kemalist regime, therefore, after having cemented the foundations of the new state, promptly launched an attack on Islamic signs and symbols. The veil for women and the fez for men were abolished in November 1925 by vote of the Great Assembly. These measures—much more than the abolition of the caliphate—provoked hostile responses from both the Turkish-speaking population and the Kurds, but this did not impede the regime the next year from replacing the Muslim calendar by the Christian calendar.[67]

With the symbolic clothing out of the way, Mustafa Kemal turned to the written language, and in 1928 the Arabic alphabet was replaced by the Latin alphabet. Thus, the break of modern Turkey with its caliphal past was carried boldly further, and a cultural rapprochement between the young republic and Europe, and with the Greco-Christian culture generally, was begun at the same stroke.

Secularity or, more precisely, the battle between secular and religious Turks, has known its ups and downs over the course of evolution of Turkish society. Four phases may be distinguished:

The first phase, the Kemalian phase, began with the proclamation of the Republic and ended in 1938, the year of Atatürk's death. This phase was marked by the establishment of an authoritarian and charismatic regime with a single party (the Republican People's Party) and by highly secular reforms.

The next phase began with World War II and continued into the postwar period. From 1945 onward, political life would be dominated by political liberalization (the establishment of bipartism) and, on the strength of the electoral interests of the newly created democratic party, by a religious liberalization. This period ended with the military *coup d'ètat* of General Gürsel, but the fundamental realities of Turkish political life remained intact. Bipartism would continue to exist and religious activities were not disturbed.[68]

A third phase is distinguishable in the late '60s and early '70s, when bipartism gave way to pluralism. In January 1970, a party representing an extremist Islamic tendency was founded by Necmettin Erbakan. This new political formation, which curiously chose the name Party of National Order (later renamed "The Party of National Welfare")[69] would later play an important role in Turkey's Islamic "renaissance." The Party of National Welfare has been represented in the National Assembly since 1970, when it won 48 of the 450 seats, quite a few for an organization newly entered upon the political scene.[70] Strengthened by their success, Erbakan and certain of his aides were members of several coalition governments in 1974 and later, but after their setback in the 1977 elections (when they obtained only 20 seats), the Muslim extremists began to incline increasingly toward a "revolutionary" solution. The Iranian events that had begun in the summer of 1977, and the rise to popular prominence of a man such as Khomeini, certainly contributed to the even more radical postures assumed by the leaders of the Party of National Welfare as they openly declared their hostility to the Kemalist reforms, and consistent to their new strategy they descended into the streets to agitate in Khomeini style.[71]

Continuing agitation, serious social and economic problems, and summary assassinations induced the heads of the army, who have traditionally seen themselves as the authentic guardians of the Kemalist heritage, to intervene militarily, opening a new phase in Turkish political life. The constitution of November 22, 1982, written under the tutelage of General Evern, intones once again the major refrains of Kemalism, and it appears that the Turkish state will accordingly remain *national* and *secular* above all else.

Of course this theoretical model does not entirely reflect the political realities of Turkish society. Clearly it is not the Turkish nation that decides its own destiny, although theoretically it may occupy the seat of authority. Its fate is decided first and foremost by the propertied classes and the army. It is they who act in its name. This being the case, the propertied classes have always sought to justify their policies in the name of the "supreme interests" of the nation. Islam does not figure in their arguments in justification and legitimation of their actions. But it may be plausibly argued that the incessant juggling act with the concept of the "nation" in the political discourse of the dominant classes has had the unfortunate result of driving the dispossessed classes to use the terms of Islam to demarcate themselves from those wielding authority over them.

To conclude, we concur with Şerif Mardin that the revival of Islam in Turkey "has been accompanied by a number of factors of a social, political, and economic nature. This complicated interplay has given rise to an idea that has never lost its attractiveness: namely *social justice*. It is because social justice *per se* is a powerful concept in our time that the idea of *Islamic* social justice has become so attractive to the Turks. Consequently, it is in social structures that the answer to the following question is to be sought: Will Turkey have a Muslim future, and if yes, to what degree?[72]

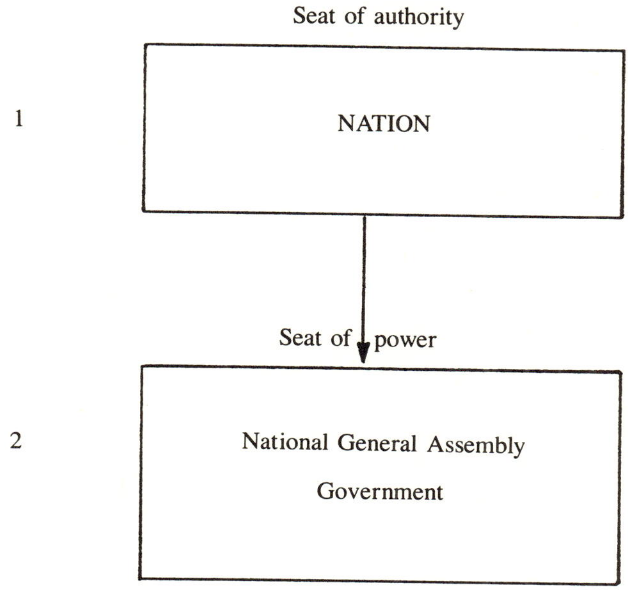

Figure 5. The Turkish Republican Model of Power

Movements of a monist nature: The Muslim Brotherhood

The Muslim Brotherhood (Ikhwān al-Muslimīn) is the best known of all contemporary Islamic movements. Not only is it older than other movements (it was founded in 1928), but there also exists an extensive literature on it in many languages.[73] Therefore there is no need to repeat here what has been said many times before. Instead, we shall concentrate on some of the guiding ideas of the Muslim Brotherhood, and endeavor to relate them to the aggregate body of doctrine of relatively similar movements.

The basic idea of the Muslim Brotherhood is the restoration of a *total and universalist Islam*—"total Islam" in the sense that the Brotherhood adheres totally to the totality of Islam. This means that in their eyes Islam is

> an all-embracing order bearing on all aspects of life: it is state and country, government and nation, creation and force, mercy and justice, culture and law, science and decree, material and wealth, acquisition and

> opulence, holy war and appeal to God, army and thought, and similarly it is true doctrine and religious worship.[74]

At this level the Muslim Brotherhood has much in common with certain contemporary Islamic movements such as Khomeiniism, which, moreover, has been strongly influenced by the Muslim Brotherhood movement—so much so in fact, that it might even be considered the Shī'ī branch of Ikhwān al-Muslimīn. The Ayatollah had formulated the same idea of a total Islam as the Muslim Brotherhood long before he came to power. He declared that "Islam is a program for all the problems of society: the form of government and administration, regulation of transactions among individuals, relations between the state and the people, relations with foreign states, and all other economic and political questions . . . "[75]

The "total Islam" of the Muslim Brotherhood is clearly different from every form of selective Islam. It is both the Koran and the *sunna*, and embraces even the opinion of a "doctor of law or his *locum tenens* on all that has not been yet dealt with in an explicit text."[76] Clearly, this aspect of the total Islam of the Muslim Brotherhood is unacceptable to partisans of a selective Islam, such as Colonel Kaddafi. Kaddafi, while allowing that the Koran is the source *par excellence* of any and all works, individual and collective, rejects the *sunna*, and even more so the utterances of the doctors of the *sharī'a*, as several sources of inspiration and legislation.

The total Islam of the Muslim Brotherhood will also differ from the Islam of the Mujāhidīn-é Khalq. Unlike the Ikhwān, the Mujāhidīn do not regard Islam as the unique and exclusive source of inspiration and action. To be sure, it remains the principal source, but it is also buttressed (or complemented) by another, namely a type of Marxism.

The Islam of the Muslim Brotherhood is a universalist Islam, in the sense that the Muslim Brotherhood avows its independence—at least in theory—of all ethnic, national, or confessional ties. It is perhaps this aspect of the Islam of the Ikhwān that is responsible for the power of its message among the Muslim population of different countries, and among different sects and social categories.

The influence of the Muslim Brotherhood in the Muslim countries where Sunnism is dominant is understandable.[77] But what is particularly striking is the success it has enjoyed in a society such as Iran, where Shi'ism has such deep roots. It is no longer a secret that the movement that overthrew the Shah's regime under the aegis of Khomeini was largely inspired by the message of the Ikhwān. A movement named

"Fedāiyīn-é Islam," modeled on the Muslim Brotherhood, was founded in Iran before World War II under the leadership of Nawwāb Ṣafavī, and supported at the time by the influential Ayatollah Kāshanī and by another Ayatollah who would become known after Khomeini's victory, Ṭāleqāni (d. September 1979). Like the Muslim Brotherhood, the Fedāiyīn made terrorist methods their specialty, and succeeded in assassinating a large number of Iranian political figures. Their bitter hostility to the government and the person of the Prime Minister Mosaddeq was in fact one of the factors that contributed to his fall.[78]

Interestingly, Ayatollah Khomeini and Sheikh Khalkhāli (who achieved a measure of notoriety through the summary executions he carried out on the orders of his master Khomeini) both belonged to the Fedāiyin-é Islam, i.e., the Shī'ī branch of the Muslim Brotherhood. Sheikh Khalkhāli is still the leader of one of the Fedāiyīn factions.

The "universalist" aspect of the Islam of the Ikhwān is also contrasted to the "particularist" and necessarily restrictive Islam preached by Kaddafism and the Mujahīdīn-é Khalq. Despite its original desire to present itself and to be received as a universalist Islamic movement, Kaddafism nonetheless has remained a very limited and particular movement. Libyan Islam, all things considered, is too personified and too identified with the person of Colonel Kaddafi to aspire to Islamic universalism. The same applies to the Mujāhidīn-é Khalq with their curious blend of nationalism and Marxism.

The foregoing discussion, as well as the fundamental ideas of the Ikhwān on the question of power, are plausible enough basis for the assertion that the Ikhwān model is directly derivable from the caliphate model, established by the dominant current of Islam (Sunnism). But although for the Ikhwān the caliphate was certainly an authentic and respectable Islamic institution, it was by no means the one, only, and eternal form of Islamic power. Here in fact is probably where the Muslim Brotherhood may be said to depart from the classic model.

To take such a position was indeed a novelty at the time their organization was founded, and required considerable political and ideological courage, considering that as a movement the Muslim Brotherhood was created only four years after the abolition of the institution of the caliphate at Istanbul. At the time, and long afterwards as well, many dreamed of its reestablishment (e.g. the Syrian Rashīd Reḍa, eminent theologian and disciple of Muḥammad 'Abduh). The Muslim Brotherhood, with a keen sense of the political realities, understood from the outset that a change was necessary, and that this change

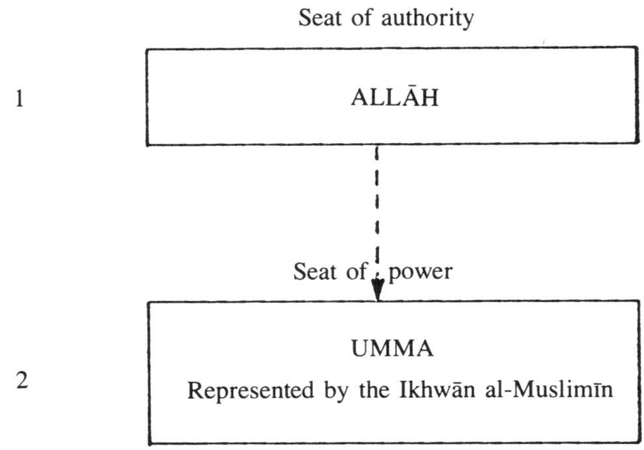

Figure 6. The Muslim Brotherhood Model of Power

would entail replacing the caliphate by the *Umma*, with its face lifted and its status regained (Figure 6).

Semi-monist models

The Algerian model (Front de la Libération Nationale (FLN)

The Algerian model can be termed *integrated*, without being *integrating*, as is evident from the 1976 national charter and constitution.

Apart from the ambiguities that arise from the very nature of relations between Islam and the state, there are two different, albeit concurrent, tendencies in Algerian constitutional institutions and declarations.

The secular tendency

The secular tendency is solemnly proclaimed in Article 1 of the Constitution, which states that "Algeria is a popular and democratic republic, one and indivisible." This wording has unmistakably secular overtones, a tendency that was further reinforced by the national charter,[79] which, after reaffirming the democratic and popular nature of Algeria, adds two other qualities, i.e., the *national* character of sovereignty and the *socialist* character of the state.

In its very first heading, "Edification of Socialist Society" the Charter attempts to reconcile the two concepts of *nation* and *people*. "Algeria," it proclaims, "is a nation," which is then defined in both its negative and positive aspects: *negative*, when the Charter stipulates that "the nation is not an assembly of people or an assemblage of disparate ethnic groups," and *positive* when it specifies that "the nation is the people itself, as an historical entity, acting consciously in everyday life within a defined territory, with a view to carrying out with *all the citizens* the tasks of a single destiny, and sharing the same trials and the same hopes."

This definition, secular in essence, leaves no doubt that sovereignty belongs to the people. Further, the motto of the republic is "by the people and for the people."

The secular nature of the Algerian state is further accentuated by the principle of equality of all citizens before the law, and equal justice for all. No discrimination among citizens is permitted. The singularity of the Algerian Constitution and Charter will undoubtedly be better appreciated if it is compared with the Iranian constitutions of 1906 and 1979. In the latter two documents, especially the 1979 Constitution, "citizens" of recognized minority confessions (Zoroastrism, Judaism, and Christianity) are treated differently from Muslim "citizens" of the *ja'far* rite (*imāmiyya*), who are favored. In other words, in this model Islam takes precedence over Iranian nationalism. This is definitely not the case in Algeria. Although the Constitution specifies that the president of the republic must be of "Muslim confession" (Article 107), there exists no distinction *in law* among citizens of different confessions. It would be no exaggeration to say that the confessional specificity of the Algerian president was comparable to that of Scandinavian monarchs.

But beyond the basic texts, what in particular reinforces the secular character of the Algerian state is the specificity of the Revolution of November 1, 1954 which ultimately gained independence in 1962. The Algerian Revolution was neither an Islamic revolution nor a revolution led by the *'Ulamā'*; it was a *national* revolution. To be sure, both Islam, as a cultural and religious factor, and the *'Ulamā'*, as an influential group of Algerians, played a major role in this revolution. But this was still a far cry from the Islamic revolution carried through in Iran under the aegis of Ayatollah Khomeini. Whereas the Algerian Revolution, under the leadership of a national political organization (FLN), was accomplished first and foremost in the name of the Algerian fatherland, the Algerian nation, and the Algerian people, the Islamic Revolution in 1979 in Iran, under the leadership of a religious caste (the *'Ulamā'*) was a revolution in the name of Islam, by Islam, and for Islam.[80]

But it must not be thought that the Algerian regime is a pure unitary model national in essence, similar to the Turkish model. Islam has a strong presence in Algerian society and in the Algerian state, and it is this presence that has led us to consider the Algerian model merely a model of the *unitary tendency* to distinguish it from *pure monist models*. Let us examine how the presence of Islam gave the Algerian model a quality all its own.

The Islamic tendency

The Islamic tendency, clearly present in the Algerian system of power, is neither strong enough to be predominant (pure monist) nor weak enough to be ignored. Yet the weight it carries does not sufficiently offset that carried by the secular side of the regime to enable the latter to be classified as dualist. In Algeria Islam occupies a place somewhere among these three variations.

According to the Charter and the Constitution (Article 2), "Islam is the religion of the state," but not because of divine will (providence), or because Islam is the best of religions, or even because Algeria intends to establish an Islamic regime like the Medina community or others; if Islam is the religion of the state it is simply because—and the National Charter specifies as much—"the Algerian people are Muslim."

Hence it may be fairly said that "the seat of authority is occupied not by Islam, but by the "Algerian people." Islam is integral to the system, but it is not itself the integrating force. Thus the integration of Islam into the system, and the fact that Islam issues from the people and its sovereignty, paradoxically reinforce the secular character of the Algerian regime.

In an outstanding article on the status of Islam in Algeria,[81] Henri Sanson provides a very good description of the intermeshing between the state and Islam. "In Algeria," he writes, "there is neither caliph, nor sultan, nor king who is invested with powers by God. There is neither a descendant of the prophet, a new prophet, a lieutenant of God, or a commander of believers, who possesses a particular authority derived from God." What then is the nature of the authority of the state? "The authority of the state in religious matters," he answers, "is

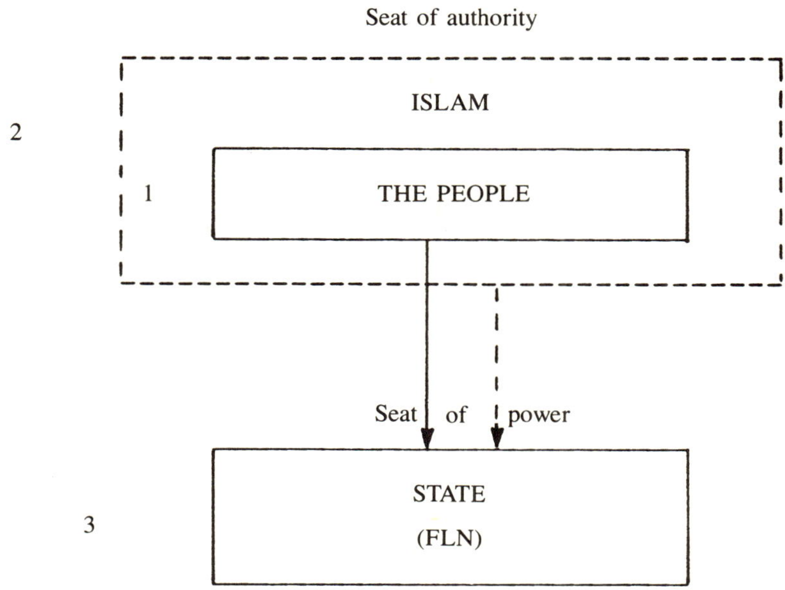

Figure 7. The Algerian Model of Power

one of *service*. Its authority could never be said to derive from divine law. Or, in other words, Islam, but not men of religion, nor the pleasure of men of government, reigns. Islam reigns, but the lay people of Islam govern.''[82]

In the National Charter, Islam is dealt with in a section following that devoted to ''Algeria: one people and one nation.'' It bears mentioning that Islam is not dealt with separately even in the section relating to it, but curiously must share the section with the ''socialist revolution.'' Thus Islam is presented as ''an integrating factor'' in the ''historical personality'' of Algerians, and is therefore more a cultural phenomenon than a source of authority or a form of state.

Islam is likewise present in public institutions; there is a Minister of Religion, a department for Koranic instruction, a Supreme Islamic Council, etc. But all these religious institutions are part of the public administration, and under the control of the state.

To sum up, the Algerian model could be described by the succinct formula that in Algeria Islam reigns, and the state ordains (Figure 7).

The Syro-Iraqi (Ba'th) model

The political system currently in power in Iraq and in Syria was inspired by the ideology of the Socialist Party of the Arab Resurrection (Ḥizb al-Ba'th al-'Arabī al-Ishtirākī). The Ba'th Party in turn took its inspiration from four different sources: Arabism, socialism, Islam, and a romantic German ideology.

The ideology of the Ba'th party is incontestably a secular ideology elaborated by secular intellectuals (Michel 'Aflaq, Ṣalāh Biṭār, and Akram Ḥurānī) and applied by secular individuals, either military or civil.

Let us attempt to examine the first three sources listed earlier, leaving the fourth aside as secondary. This should give us a clear understanding of the links between the three principal constituents of the Syro-Iraqian political system and at the same time delimit the specificity of the seat of authority in the system.

Arabism

Arabism is the principal element and guiding idea of the Ba'th ideology. The founding ideologues of the Ba'th party certainly did not believe in any nationalism specific to the present-day Arab geographical entities. They believed that this type of "chauvinistic and limited" nationalism proper to each of the Arab countries individually was only a primitive form, outmoded by Arab nationalism. Michel 'Aflaq commented on this point as follows in 1955: "I remember that twenty-five years ago in Syria one rarely heard the leaders utter the words *Arab* and *Arabism*. Rather they used the words *Syria* and *Syrians* to extricate themselves from numerous impasses created by nationalist utterances, but used patriotic expressions mainly with the aim of gathering together everyone (Moslems, Christians, Arabs, and non-Arab minorities) in the name of patriotism and to resist the foreign occupier."[83] 'Aflaq's judgment of Egyptian nationalism (*Egyptianism*, we might say) is even harsher. "In Egypt," he writes, "the problem is posed in different terms, since the Egyptian people are composed of a single race. But after so many years under foreign domination, the Turks and the English diluted the Arab character of its people, reinforced a regional sentiment, and created a certain Egyptian sense of pride."[84]

Thus it is clear that the nationalism preached by the Ba'th leaders is conceivable only in its finite form: *Arabism*. Arabism "is the primary driving force of Arabs in this stage of their life.'' Nationalism embraces in its fold all Arabs and is ''the only motto that can strike at the hearts of Arabs, penetrate to the depths of their being, and respond to their true and essential needs.'' Finally ''Arabs, their liberty, their sovereignty, and their unity mutilated, can understand no other language than the language of nationalism.''[85]

The path to Arabism conceived by the Ba'th Party is socialism, a socialism that befits the identity and meets the needs of Arabism.

Arab socialism

For the Ba'th Party socialism is an absolutely indispensable element and a *sine qua non* condition for the ultimate fulfillment of Arabism.

> We can say that Arab nationalism is synonymous with socialism in the present era and that there is no contradiction, antagonism, or struggle between nationalists and socialists. The Arab nationalists understand that socialism is the surest way to effect a renaissance of their nationalism and their nation because they know that the struggle of Arabs in the present period depends on all Arabs, and that they cannot take part together in this struggle if they are exploited and divided into masters and slaves.[86]

Not only is Arabism inconceivable without socialism, Arab unity can only be achieved through socialism. ''Unity does not consist in political actions, negotiations, or agreements concluded between governments, it is a revolutionary action and a struggle managed by the people alone.''[87] Further, ''our linking of unity to socialism is neither caprice nor improvisation; we have merely discovered that this is the only solution which will make of unity a living and moving reality in our lives.''[88]

Socialism is a necessary condition for Arabism, and furthermore its ''only true guarantee,'' because ''the definition of Arabism is not sufficient unto itself. The essential is that we have defined our nationalism through the idea of freedom as well,'' explains Michel 'Aflaq. ''The interdependence of nationalism and socialism is the *de facto* and true guarantee of Arabism.''[89]

The socialism of the Ba'th Party has its own distinctive features. Its *Arab* quality is meant to distinguish it from *international* socialism:

> We do not preach internationalism and therein lies the fundamental difference between us and the communists. We think that the human feeling in

> communists is false. Their internationalism is only an attempt to efface the personality of every people and establish artificial ties of a purely economic nature among everyone.[90]

There is also another reason adduced to justify the specific national character of Arab socialism, namely the history and general social condition of Arabs. Comparing Arab socialism to communism, 'Aflaq makes the following observations:

> Communism is the product of European thought; it is born of the situation of Europe in the modern era. For them (the Europeans) it represents the culmination of the humanist movement that appeared in Europe in the sixteenth century. Communism is also the product of the eighteenth century, with its abstract rationalist philosophy, of which the principles of the French Revolution are the consummate expression. Finally, communism emerged as a consequence of the prodigious development of industry in the last century, with its diverse economic theories and social doctrines." Thus "the Europeans, whatever the country, whatever their social or political conceptions, do not feel the same degree of aversion to communism as do the Arabs, inasmuch as there are neither connections nor kinship between communism and Arab history, or Arab life past and present.[91]

Summing up, one could say that the Ba'th are endeavoring to achieve not "Arab socialism" but "social Arabism," which is by no means the same thing.

The place of Islam in the Ba'th ideology

There is no doubt about the secular character of the Ba'th Party. An Iraqi Ba'th theoretician has made the following coment in this regard: "The Ba'th is a nationalist movement that speaks to *all Arabs*, whatever their religion or confession, considers freedom of belief sacred, and respects all religions equally."[92] But the Ba'th is not an anti-Islamic movement. Quite the contrary, it sets a high value on Islam. But what, then, it may legitimately be asked, is Islam for the Ba'th Party? Is Islam a religion of universal calling? Or is Islam a religious state? The writings of the ideologues leave no doubt that it is neither the one nor the other.

For the Ba'th Islam is an *Arab* cultural good, a "civil religion," to use Durkheim's expression. Islam was the means whereby Arabs in the past were able to compose themselves and enter history to achieve their renaissance. In other words, Islam was in the past what Arab nationalism is today. Further, Islam represented the Arab resurrection of the past, while Ba'th incarnates the Arab resurrection of our own day. In the past, "the Arabs were called upon to believe in a single god and this faith enabled them to achieve the economic and social revolution they needed. . . . Today, on the other hand, the prime driving force of Arabs in this stage of their life is nationalism."[93] Just as, as we have seen, Ba'th ascribes no sacred character to *its* Islam, it also places no stock in either the justice or utility of a religious state, whether Islamic or otherwise. "The experience of a religious state," writes Michel 'Aflaq, "had already been attempted in the Middle Ages. But it ended inauspiciously, costing humanity much travail and much blood. The same attempt failed in the Moslem East as well. The result, known to one and all, was a steady succession of wars and revolutions."[94]

The secular approach of the Ba'th has obviously not been without its repercussions on the constitutions and institutions of Iraq and Syria. But even despite the considerable efforts in this regard, there still exist major spots of resistance, and attempts to repudiate the sociopolitical project of the Ba'th are numerous. In the face of these obstacles, the Ba'thists, further weakened by serious internal discord and conflicts, have been forced to adopt a more flexible and even conciliatory attitude toward Islamic tendencies. Recent events in Iraq and Syria have also had quite negative repercussions on the secular aspect of the Ba'th, especially in Iraq by reason of the war with the Islamic regime in Iran (not without its paradoxes), and in Syria, because of the growing influence of Moslem zealots.

Moreover, an analysis of the texts of the Syrian and Iraqi constitutions shows clearly the limits encountered by a secular and secularizing policy in societies in which the Moslem religion still exercises an influence as varied as it is deep on the minds and mores of a large segment of the population.

The privileged status accorded by the Ba'th to Islam appears in both theoretical writings and institutional texts. For example, although the statement that "Islam is the state religion" does not appear in later Syrian constitutions, the constitutions of 1950, 1953, and 1964, and finally the "definitive" Constitution of 1973, state that "Islam is the religion of the President of the Republic" and that "Moslem law is the principal source of law." The same principle had been affirmed earlier in the 1969 Constitution (Article 3). Indeed, the 1969 Constitution, written when General Ṣalāḥ Jadīd (a Ba'th "hardliner") held power,

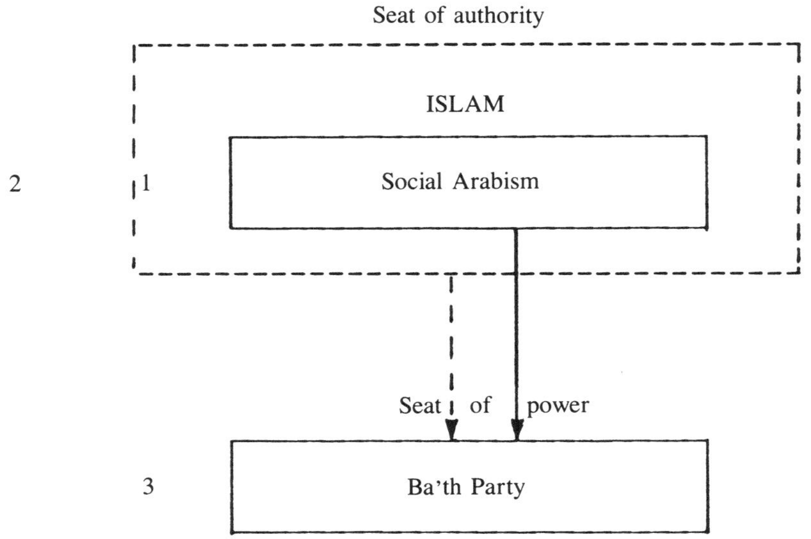

Figure 8. The Syro-Iraqi Model of Power

resembled the constitutions of the socialist countries in certain respects, and can be considered a genuine step toward the secularization of power inasmuch as the article concerning the religion of the President of the Republic disappeared and the oath "I swear in the name of Allāh" was replaced by "I swear on my honor and my faith to act and to struggle for the achievement of the aspirations of the Arab nation toward unity, liberty, and socialism" (Article 59).

In reintroducing the article on the religion of the President of the Republic in the 1973 Constitution, General Hāfiẓ al-Assad had wished to appease religious circles. But the constitution does not state that Islam is the state religion. In contrast, the provisional Iraqi constitution of July 27, 1958 (promulgated under General Qāsim) and the provisional constitution of September 28, 1968 (promulgated under General Al-Bakr) proclaimed that "Islam is the state religion" (Article 4 in the two constitutions) just as does the Constitution of July 16, 1970. On the other hand the Charter of National Action, proclaimed November 15, 1971 by General Al-Bakr, which was intended to serve as a basis for drawing up a "permanent" constitution, contains no reference to the relationship between the state and Islam.[95]

The conclusion from this would seem to be that the Ba'th model is a typical example of elitist regimes (civil or military) of a clearly secular calling, that must, however, deal with confessional obstacles which for the time being seem difficult to overcome. While the regime in power seems to have resolutely set its sights on *civil unity*, the traditional social structures, on which religion exercises a profound influence, stubbornly resist this tendency.

All these things considered, compared with other models currently in force in the Moslem countries, the Ba'th model may be safely said to be the most secular (outside of Turkey). It is much closer to the Destur-ien model in Tunisia than that of the FLN in Algeria (Figure 8).

Dualist models

The Saudi model

It may be surprising to see the Saudi regime classified as a dualist model, since it may validly be asked how a regime that has chosen the Koran as its one and only constitution could be considered "dualist," and in fact, from this perspective the Saudi model could be considered a monist model of religious nature, comparable to the Khomeini model. But for the comparison to hold, the channels for conveying authority would have to be delineated as clearly and precisely as they are in the Khomeini system. The mere fact that a regime refers to the authority of the Koran is not in itself sufficient to categorize it as religious and monist. It is also necessary to determine the mode whereby legitimacy is conferred from the preestablished source. An examination of the classification of models presented in the foregoing will show that the modes for conferring legitimacy in Islam are few. In the traditional model, there are three known modes, i.e., the *caliphate*, the *imāmate*, and the *Umma*. The modern "Islamic" models tend more to notions such as the *nation*, the *people*, or a fusion of the people with God. But where does the Saudi model fit in this list?

The Saudi model resembles neither the caliphate nor the imāmate. As for the *Umma*, it is noteworthy that the Saudi regime has always taken pains not to regard it as the source of its legitimacy.

As for the modern modes, the Saudi regime has never claimed to be a *national state*, nor does the nation in the proper sense even exist in Saudi Arabia. Moreover, the terms "nation" and "Saudiism" are contradictory by definition and cancel one another out.

With the "nation" eliminated, can it be replaced by the "people"? Would it be precise to regard the "people" as the seat of Saudi legitimacy? Some authors think so. Edward Mortimer writes that "Saudi

Arabia is not a theocracy but an *Islamic monarchy* in which the king draws his authority from the people.''[96] While wishing to avoid a discussion on the definition of ''people,''[97] it might nonetheless be asked by what mode authority is conferred by the people onto the king. There would have to be evidence that the people had some way to pronounce themselves on the matter. But in a country like Saudi Arabia where elections are totally unknown, the people are deprived, practically speaking, of any form of expression. That alone would preclude ranking the Saudi regime among those deriving their authority from the people.

One possibility remains open, then: a fusion between the people and God—i.e., the fusionist model, which is the model of Kaddafi and the Mujāhidīn-é Khalq; it first of all requires the elimination of the intermediate seat (the *'Ulamā'*), but the Saudi regime does not in fact meet this condition. As a source of religious authority and an influential group, the *'Ulamā* enjoy a rather strong presence in Saudi life and the Saudi political system. The Saudi religious caste, moreover, has the singular feature of being made up mainly of the family of Muḥammad b. 'Abd al-Wahhāb, today known under the name of Āl al-Shaykh. As will be seen further on, this caste had a large hand in building the Saudi state and is today one of its constituents. Hence, because of the active presence of the *'Ulamā'*, the Saudi regime cannot be classified as a fusionist regime.

Having eliminated these possibilities, the task remains of determining the true character of the Saudi model, and that would seem to be a dualist model whose origins go back directly to the founding act of the Saudi state. Indeed, the Saudi state is essentially the product of the famous alliance of Dar'iyya (concluded 1744 A.D.) between the celebrated Muḥammad b. 'Abd al-Wahhāb (1703–1792) and the emir Muḥammad b. Sa'ūd, from the powerful confederation of tribes, the 'Anaza (or 'Unaiza).[98]

Since then, the basic facts of Saudi power have scarcely changed. On the one hand, there is Islam, incontestably the prime constituent of the regime; secondly there is tribalism, which can boast of a strong and active presence in the system. The relations between the two are dynamic, marked by conflict and cooperation. The following analysis is a study of these two constituents of Saudi power.

The Islamic component of Saudi power

The Islam of Saudi power is Sunnī Islam of the Ḥanbalī rite, in its Wahhābī variant. Saudism has also been called neo-Ḥanbalism; it has steadfastly refused to give exclusive veneration to Ibn Taymiyya and his great disciple Ibn Quāīm al-Jawziyya.[99]

Ibn Taymiyya was one of the most intransigent jurists as regards the strict application of the law (*sharī'a*), and elaborated his political theory in three famous works.[100] That theory may be summed up in the following cardinal points:

1. A recognition of the legitimacy of the prophet's succession. Essentially this meant acceptance of the caliphate of the Four Orthodox Caliphs in the order in which they actually came to power.

2. The Community's need for political power for the community. (Ibn Taymiyya's view on this point was mentioned above).

3. Repudiation of the uniqueness and universality of the caliphate. On this point Ibn Taymiyya's view clearly diverges from the dominant position, which emphatically affirms those qualities. On the other hand, Ibn Taymiyya's political pluralism bears a certain kinship with Khārijism.

Based on the two preceding points, it could legitimately be said that points 2 and 3 contain the first germs of the future Saudi state.

4. Rejection of the practice of the *'Ulamā'* electing the head of state. Ibn Taymiyya points out that essentially "there has never been, even at the very beginning, a true election, but rather a quite rapid and complete concurrence of the people with a designated or imposed head of state."[101]

On this point as well, Taymiyya differs from the majority of Sunnī judges and theoreticians. But in Saudi Arabia his view has only been partially applied.

5. A lack of severity and a tempered demand for the requisite qualities of the Prince of Muslims. This question, of course, has been the subject of permanent debate, and discord as well, among Muslim theoreticians. While the "idealists" or "maximalists" require supreme physical, intellectual, and moral qualities, sometimes approaching even the superhuman, of a candidate for the supreme magistrature, the "realists," Ibn Taymiyya included, were satisfied simply with the moral qualities required of a person in order to accept his witness.[102]

The essential point is that Ibn Taymiyya's flexibility on this issue seems quite probably to have played an important role in the decision taken by his distant disciple Muḥammad 'Abd al-Wahhāb to strike up an alliance with the emir Sa'ūd.

Muḥammad b. 'Abd al-Wahhāb drew on the political and legal theory of Ibn Taymiyya, itself derived from the teachings of Ibn Ḥanbal, and adroitly turned the ideas of his intellectual masters to account in action. Native of Najad, Al-Wahhāb was educated mainly at Medina and in Iraq, taught at Cairo as well as at Qom (the sacred Iranian city) and was the author of the "Treatise on Oneness" (*kitāb al-tawḥīd*).[103] The model we elaborated fits nicely with both the historical past and the current circumstances of Hidjaz. Laoust describes Wahhābism as "an Arab movement of political and religious renewal, with a view to organizing a state in accordance with the principles of public law as defined by *Siyāsa al-shar'iyya*, at a time when the Ottoman Empire was beginning to show its first signs of disintegration."[104]

This concise and thorough definition helps to place the Wahhābī movement in its historical context. At the time when the theologian Ibn Wahhāb and Amir Ibn Sa'ūd concluded their alliance, Arabia was a part of the Ottoman Empire dominated by the Muslim faith. Or, in terms of the traditional model, power was wielded by the Ottoman caliph-sultan. The intermediate sphere, usually occupied by the *'Ulamā'*, was practically speaking vacant, since under the Ottomans the *'Ulamā'* had ceased to represent what they had represented formerly in the Umayyad and 'Abbāsid empires.

The emir Ibn Sa'ūd, who aspired to be more than a mere provincial governor appointed and controlled by the central administration, had no other choice in this situation than to abandon the traditional model if he wished to pursue his ambitions unimpeded. It might be objected that he could have done the same as the Būyid emirs or the Seljūq sultans had done before him. But there is a difference in dimension that distinguishes the case of Ibn Sa'ūd from the others: unlike the Būyids and the Seljūqs, who disposed over considerable military forces, sometimes superior even to those of the caliph himself, the Amir had no more than a small body of barely armed warriors. This made all the difference between their several postures. Moreover, the Būyid emirs and Seljuq sultans, like indeed most of those of varying ethnic origin who had taken up the 'Abbāsid challenge (for example, the Tahirids and the Ghaznavids), were not Arab, and had to confront a caliphate that had linked its legitimacy to its Arab ethnicity. The result was that the non-Arab emirates and sultans, despite their formidable power, were forced by necessity to recognize the legitimacy of the caliph in power and *de facto* to come to terms with him. The Saudi case was completely the reverse. This time it was a non-Arab caliph, a Turk, who was challenged by an Arab pretender, a native of the land of Islam's birth.

The non-Arab ethnicity of the Ottomans could have been an issue in the later revolts of the Saudis, but it was not. Muḥammad b. Sa'ūd and his descendants took great care not to play this hand and not to challenge the foundations of the Ottoman caliphate. The reason was that the Saudis, adopting the ideas of Ibn Taymiyya, attached more importance to political pluralism than to the oneness and universality of the caliphate. They were therefore able to find arguments allowing them simultaneously to acknowledge the Ottoman caliphate politically and juridically, on the one hand and, on the other, to reserve the right to constitute their own state. This was in the end the practical result of Ibn Taymiyya's political realism. Later, that realism would be translated into facts by Ibn Sa'ūd and Ibn Wahhāb.

The abandonment of the traditional model of Islamic power required its replacement by another model, which in turn required the active hand of a reformist theologian with the necessary acumen for this novel undertaking and at the same time prepared to vouch for his own efforts as well.

Finally, the new model, once created and underwritten, required a driving force for its implementation. It was at this stage that Muḥammad b. Sa'ūd, tribal chieftain, intervened. The historical record shows that the Saudis placed all their forces and all their means at the disposal of the project of the theologian Ibn Wahhāb from the very outset. In an act of alliance (*bay'a*) the two men pledged themselves to effecting the triumph of the word of God, by force of arms if necessary. The official emblem of Saudi Arabia, made up of two sabers (symbol of power) enlaced by the first formula of *shahāda* (the majestic sign of Islam) is undoubtedly a living memory of the Wahhāb-Sa'ūd alliance of 1744.

This alliance required almost two centuries of strenuous effort, marked by cruelties, wars, and pogroms, to arrive at the establishment of the Saudi state In 1932, Amir 'Abd al-'Azīz b. Sa'ūd succeeded in bringing the different provinces of Arabia under the same banner. On October 13, in the same year in which the Ottoman caliphate was abolished (1924), 'Abd al-'Azīz made his triumphal entry into Mecca and proclaimed himself the "guardian of the holy places."

The assumption of such a title at the very moment the caliphate had become vacant might have encouraged the emir to proclaim himself caliph. But motivated by a keen political realism and above all faithful to the teachings of Al-Wahhāb, the emir refused to take this step, effectively repudiating the traditional model. However, the same cannot be said for his Hāshimī rival, Shārif Ḥusayn, who, although (or perhaps because) he was losing on all fronts, unhesitatingly proclaimed himself caliph in the same year, 1924.

Interestingly, after their alliance with Wahhābism the Saudi emirs consistently avoided all titles having a religious connotation. They claim to be neither caliph, imām, nor prince of believers (*Amīr al-Mu'minīn*), their modesty in this regard contrasting with the temerity of Muslim sovereigns in certain other Islamic countries: for instance, the title of "Imām" heads the long list of Khomeini's titles; and the king of Morocco still keeps his title of Amir al-Mu'minīn.

The Saudis do not even claim their power derives from providence (*mashīyya*), which strongly suggests that the Saudi regime is in fact not a typical theocracy. The sources of its legitimacy are two: *Saudism* and *Wahhābism*.

Like any other religious movement, Wahhābism has undergone a number of different interpretations, with two currents in particular predominating: the *puritanist* current and the *official* current. The first spawned the Ikhwān movement. The Ikhwān believe that Wahhābism should call for a definitive return to the pristine past of Islam—i.e., the epoch when all Muslims were linked by the true bonds of fraternity (*ukhuwwat*), in an allusion to the Medina community where the partisans of the new religion became "brothers" on the Prophet's order. As the Koran says, "Believers are merely brothers. Therefore establish concord among your brothers and be pious toward Allāh. Perhaps he will be merciful unto you." In brief, the Ikhwān, emulating the act of the first partisans of the prophet (those who went out—*muhājirūn*) departed for *hijra* to establish the Medina model in the form of agrarian communities (Islamic collective farms). Their categorical rejection of all forms of modernity (telephone, radio, taxes, etc.) brought them into conflict with the Amir 'Abd al-'Azīz, who as a political leader had intended to build a "state" and not a "community." From 1924 to 1932, armed clashes and conflicts between the two parties multiplied. Finally, unable to stem the Saudis' irresistible rise to power, the Ikhwān were forced to become part of the Saudi state. They were later appointed guardians of public morality, the merciless executors (*mawtaw'a*) of the formidable institution called the "recommendation of good and prohibition of evil" (*amr bil-Ma'rūf wa nahy min al-munkar*). But more recent events have shown that the assimilation of the Ikhwān was not as successful as some would have liked to believe. The rebels who occupied the Great Mosque in Mecca on November 20, 1979 had symbolically chosen the name of Ikhwān. Their leader, the young Juhaym al-'Utayba, was in fact a relative of one of the Ikhwān leaders of the anti-Saudi uprisings in 1928–1930.

Counterposed to the Ikhwān is official Wahhābism, represented by the descendants of Al-Wahhāb, who give the regime valuable moral support, at the same time imparting to it a certain "Islamic authenticity." Enjoying numerous privileges, respected, and even participating in the government, the Āl al-Shaykh serve in practical terms as the cohesive element to the regime. They are also linked to the Saudi family by bonds of kinship: the powerful Saudi king Fayṣal was a descendant of Muḥammad b. 'Abd al-Wahhāb on his maternal side.

The Āl al-Shaykh are active at two levels in the affairs of the kingdom: in regulating and interpreting the *sharī'a*, and in the taking of political decisions.

At the first level, the role of the Āl al-Shaykh is similar to that of their counterparts in countries where the *sharī'a* rules the law. "They of course do not represent a unified juridical estate, but their *fatwā* alone are sufficient for them to effect a credible interpretation of the *sharī'a*."[105] At the political level, their influence is perhaps debatable. Although they are permanently represented in the Cabinet of Ministers (three to four seats), it is clear that political initiatives, at least on important issues, come from elsewhere, namely the influential members of the Saudi family. Their most conspicuous intervention was in June 1964, when the *'Ulamā'*, putting their heads together, published their *fatwā*, which proclaimed the "inability of King Sa'ūd to reign." It would be simplistic to believe that the *'Ulamā'* were acting independently in doing this. King Sa'ūd had for all practical purposes been eliminated even before the publication of this *fatwā*, and real power was in the hands of his brother and prime minister, the formidable Fayṣal. The *'Ulamā'* were also directly involved in 1979 in the incident at the Great Mosque and issued a *fatwā* authorizing Saudi authorities to intervene militarily to put an end to the insurrection.[106]

Although the Saudi *'Ulamā'* do not have political autonomy, this does not mean that they are mere recorders of the sovereign will. They are, however, an integral part of the regime, and one of the two elements in the Saudi *symbiosis*, the other element being the Saudi family, which has preserved its essentially tribal structure down to the present day.

The tribal component of Saudi power

Although Islam is the generative force in the Saudi regime, the tribal character of the regime is particular to it. This concurs with the model proposed by Ibn Khaldūn in his description of the relations between religion and the tribe. At bottom, Ibn Khaldūn gives priority to real power (*showka*) and domination, with religion intervening only to ratify and justify a *de facto* situation. "Only he who dominates a nation or a race is capable of directing its affairs. Religious law will not be at variance with the facts. And God on high knows this very well."[107]

In Saudi Arabia, the "fact" that religion does not contradict is the *esprit de corps* (*'aṣabiyya*) that unites the members of the Saudi family. The clan spirit reinforces the ardor of the tribal members and "makes them fearless, for each of them puts his family and clan spirit above everything. God has placed in the hearts of men a natural affection for his own blood: it is the source of mutual succor (*ta'āḍud*) and it intensifies the enemy's fear."[108]

Further, according to Ibn Khaldūn, the clan spirit inspires a sense of racial superiority (*ghalaba*) it is not an end in itself; its aim is to establish royal power. The clan alone, however, is incapable of establishing a monarchy. To do this, it is necessary to obtain through force or consent a pledge of alliance from the other classes in one's own tribe. In the event, the dynasty "embraces the strongest clans among its clientele to establish a monarchy that is no longer absolute."[109]

Once founded, the tribal dynasty runs a heavy risk of becoming the object of conflicts and rivalries among the various clans composing it. How then can its disintegration be avoided? Ibn Khaldūn pushes his analysis to the limit, saying that only a religious structure can "efface competition and envy among those who are united by blood ties, and nothing can stop them because they are united in their perspectives."[110]

If we now turn our attention to the course of events that led to the founding of the Saudi state, from the conclusion of the *Dar'iyya* pact to the effective establishment of the monarchy, we will see that this process followed quite faithfully the path mapped out by Khaldūn's model. The following facts are persuasive enough evidence in this regard.

First, the Saudi dynasty is the product of a tribal confederation (*'anaza*). The solidarity and *esprit de corps* reigning among the various constituents of this confederation and the members of the present Saudi family are quite impressive in many respects.

To be sure, like any other living social entity, the Saudi family has not been immune to internal conflict and dissension over the course of two centuries. For example, there were the dynastic quarrels of the 1840s among the sons of the Amir Fayṣal. Similar events have marked more recent history, for example, the departure of Prince Ṭalāl in exile to Cairo in 1962 to create the Arabian Liberation Front, the deposal of King Sa'ūd in 1964, or the assassination of King Fayṣal by one of his young relatives in 1975. In all these events, what stands out is that in each instance the Saudi family was able quite swiftly to overcome its momentary divisions.

The other factor of importance is the effective participation of the various notable clans in the exercise of power. The three most representative are the *Jilwa*, to which the late kings Sa'ūd and Khāled (d. 1982) belonged, the *Shamar*, represented by the present hereditary Prince 'Abdallah and Muḥammad (Amir of Medina), and finally the *Sudayrī*, who produced King Fahd and his brothers Sultan, Nā'yīf, Turki, Aḥmad, and 'Abd al-Raḥmān.

All in all, however, King 'Abd al-'Azīz (father of all these princes and kings and many others) contributed nothing new as regards integration by marriage; this was an ancient practice, which the Prophet himself had effectively employed.

However that may be, tribal affiliation, especially in societies with tribal structures, has traditionally been one of the most efficacious ways to consolidate the base of power and ensure its persistence.

Heredity is of course a related factor in the consolidation of power. Qurayshī descent was indispensable for access to the supreme magistrature, at least until 1258 A.D., the date of the fall of the 'Abbāsids. This principle also applies to the Shī'ī, who believe in the legitimacy of the family of the Prophet (*Ahl al-Bayt*). For the Saudis, the situation is basically the same, except that Qurayshī descent has been replaced by Saudi descent. Thus power necessarily devolves to members of this family. Accordingly, Saudi tribalism would be a constitutive factor of Saudi legitimacy.

But an ambiguity remains: is this really true, or is tribalism merely a condition for access to royalty, as in the case of the Umayyad and 'Abbāsid caliphs? In the traditional model, Qurayshī descent was only a condition, whereas theoretically the caliphate derived its legitimacy from its quality of vicariate of the Prophet. Saudi royalty differs from the caliphate of the earlier era on precisely this point, and hence there is nothing *Islamic* in Saudi royality as such. One could even go so far as to

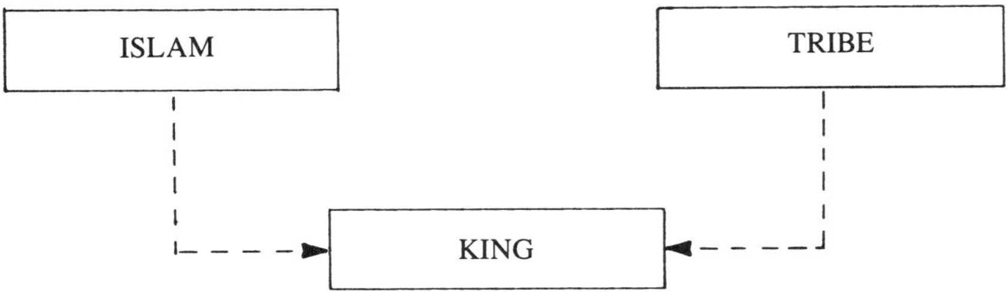

Figure 9. The Saudi Model of Power

say that this royalty is a royalty stripped bare, i.e., divested of myth and detached from providence. Though it may be true that Saudi Arabia has no other constitution than the Koran, this does not mean that its regime is more Islamic than those of Muslim countries that have adopted constitutions other than the Koran. The real reason for the absence of a secular constitution in Saudi Arabia lies in its tribalism. As a general rule, tribal or semitribal systems do not wear a constitutional regime well, and this may be found to apply as well to Kuwait (where the constitutional regime has been suspended) and indeed all the Arab emirates in the Persian Gulf. In those countries where tribalism is dominant and power is in the hands of tribal chieftains, society is administered solely by "tribal laws," which regulate daily life as well. The codification of these "laws" in a constitution would be possible only at the price of demolishing tribal structures, since tribalism obeys only its own "laws," i.e., the totality of customs that must by nature be given free play if they are to retain their vitality. Codification would lead ineluctably to ossification.

But if tribalism cannot suffer a reduction to constitutional form, it is just as unable to accommodate to a borrowed form of constitutionality based on "civil society," since political civility necessarily implies an equality of all citizens independent of ethnic social or religious factors, and is the absolute antipode to tribal values.

The absence of a constitution strengthens further the tribal nature of the Saudi regime, and it is this tribalism, ever present and ever effective, that warrants classifying the Saudi regime as dualist (Figure 9).

Fusionist models

The Libyan model

The model adopted by the Libyan Popular Socialist Arab Jamāhīrīyya does not lack originality. It is a model that claims authentic Islam as its authority[111] and is distinguished from other "Islamic" models by a number of features, including the following.

An appeal to the universality of Islam

At the secular level, Libyan Islam, inspired by the thought of Mu'ammar al-Kaddafi, claims to be the *Third Universal Theory*, valid for Muslim and non-Muslim societies alike. Kaddafi's *Green Book* presents (in his words) "the *definitive* theoretical solution to the problem of the machinery of government." He goes on to add that: "Islam is *internationalist*, *universalist*, and common to all mankind. It is the source of inspiration of ideologies, which are more comprehensive than any theory."[112]

This being so, for Kaddafi the universalism of Islam assumes its full meaning at the Muslim *level*. At this level, Libyan Islam is incontestably the most antisectarian existing today. It aspires to be an "Islam without rites nor sects struggling against one another . . . ." Everything seems to indicate that the Libyan leader is inspired by the belief that the only way to achieve Muslim unity is through the dissolution of all the sects (Sunnī and Shī'ī in particular) and all the rites (Ḥanafī, Mālikī, Shāfi'ī, and Ḥanbalī). And indeed, these sects and rites, which have undoubtedly been divisive among Muslims throughout history, belong definitively to the past. Historical conflicts must not become the sources of new conflicts among Muslims, all the more so that these sects and rites have no religious basis. "These sects (Madhāhib) began as political movements aimed at seizing power and forming a government . . . It was therefore not a religious problem, but rather a purely secular question with which they had to deal, namely that of forming a state or a group of men with particular beliefs."

The substitution of positive law by natural law

Furthermore, according to Kaddafi, Muslim unity can be achieved only if positive law (*fiqh*) gives way to natural law, i.e. the Koran and custom. "Any other endeavor beyond these two sources is useless and illogical." Kaddafi condemns *fiqh* as the "juridical" factor responsible for Muslim division. It must therefore be rejected in favor of the Koran, which is recognized as the "supreme law, the *nomos*, and the *sharī'a* of society'' (Declaration of the Establishment of the Power of the People, March 2, 1977, Article 2).

This declaration confirms the specificity of Libyan Islam. To be sure, in a number of Muslim countries the *sharī'a* is the effective source of all laws. But the *sharī'a* by no means eliminates the *fiqh*. Quite the contrary, the *fiqh* is one of the components of the *sharī'a* alongside the Koran and the *sunna*. Kaddafism accepts only the Koran, rejects the *fiqh* totally, and the *sunna* partially. It gives no credence to the *ḥadīth*, i.e. the verbal *sunna* (the words of the Prophet). On the other hand, it attaches some importance to the *sunna* action. Why this distinction in the way the *sunna* is received? Al-Assiouty, one of the interpreters of Kaddafi's thoughts, gives us the answer. He tells us that for the Leader of the Libyan revolution, the *sunna* is the *ṭarīqa* (method or path), which cannot be the case with the *ḥadīth*, since it is contradictory and therefore generates doubt—all the more so as the compilation of the *ḥadīth* was begun two centuries after the death of the Prophet, and is based on the statements of storytellers who transmitted the texts of the *ḥadīth* orally for over two centuries. For these reasons alone, the *ḥadīth* cannot serve as the law for society.

As for the *fiqh*, it must be rejected by virtue of the fact that it is the work of people, some of whom are infidels and others who are heterodox. It is also the product of a bitter struggle among the Shī'ī, the Khārijī, esoteric rites, and divided Muslims. Furthermore, the *fiqh* is the source of rites that contradict one another. Although it is incontestably a part of the Muslim heritage, it is in no way able to serve as a *sharī'a* for society. In brief, the *fiqh* may not have a higher status than, for instance, Roman law or the Napoleonic code. ''Although magnificent, elaborated by Muslims, of Islamic culture and Muslim orientation, it is a positive law.'' The only law that is not positive is the Koran, and to the Koran it is necessary to return, says Kaddafi.

But why only the Koran? The Libyan leader answers in substance that the Koran, as distinct from the *fiqh*, enjoys the unanimous acceptance of Muslims. The only book on religion is the Koran. It contains the direct solution to problems here on earth. It also contains general prescriptions for conducting one's life. Moreover, the Koran is unchangeable, not subject to revision as are constitutions. It is not the work of human beings, does not depend on passions, transcends time and space, and extends beyond national boundaries. Consequently, natural law, i.e. religion, ''which absorbs custom,'' must regain the place that positive law has usurped from it.

The abolition of the intermediary seat

The abolition of the intermediary seat is the direct consequence of the abandonment of the *fiqh*. Since it was the *'Ulamā'/fuqah'ā* who "manufactured" the *fiqh*, the abandonment of the latter will result perforce in the elimination of the *'Ulamā'* from the domain of power.

But in addition to their normal function as jurists (*faqih*), the *'Ulamā'* also serve as interpreters (mufassirū*n*) of the Koran, a function that could gain for them a sizable share of power. However, the Libyan regime does not allow them this latter function. Mu'ammar al-Kaddafi argues that the Koran has no need for interpreters; it is its own interpreter. Furthermore, "the Koran is written in *Arabic*; we can understand it ourselves without needing an Imām to interpret it for us."[113]

In addition to these *Islamic* arguments which, according to Kaddafi, justify the elimination of the *'Ulamā'*, there are other arguments of a *political nature*, of which we may mention two. The first derives from Kaddafi's basic conception of democracy and will be dealt with in the next section, "The Fusion of Allāh with the People." The second is inherent to the political substance of the Libyan view of power. There should be no doubts that the Libyan regime is first and last a military regime led by a military caste. It is only natural that the Libyan officers, who achieved power without the assistance of the *'Ulamā'*, should not want to share power with anyone, especially the *'Ulamā'*. Furthermore, it should not be forgotten that the *'Ulamā'* constituted one of the pillars of the former regime, and that King Idris presented himself both as Chief of State and the highest religious authority in the country. In this latter quality, the king represented the clergy, i.e. the *'Ulamā'*. It is therefore not unthinkable that the 1969 coup d'état was to a certain extent a move against the Libyan *'Ulamā'* as well. It was only logical that having triumphed over the former regime, the military caste would use ideological and political means to weaken if not to eliminate the Libyan religious "caste" from the sphere of power.

The mode by which the Libyan officers overthrew the Idris regime made it unnecessary to obtain legitimation via the *'Ulamā'*. Indeed, the group around Kaddafi acted alone in overthrowing the regime in power. The *coup d'état* was conceived, planned, and carried out by officers acting totally autonomously. At no time did they solicit the complicity of religious forces at any level. Thus, having both taken initiative and decisions autonomously, the officers could feel independent of other political and social groups: they bore them no debt, nor were they burdened by any compromising commitment. This *de facto* freedom, especially in a country rich in natural resources, but relatively poor in human resources, made them fully free to choose their political course. It was against this extraordinary political background, out of the ordinary, with no dynamic or organized religious form present, that Islam, as defined by a military caste, became an Islam without the substitution factor, i.e. without the *'Ulamā'*.

With regard to the traditional model of Islamic power, we can say that in today's Libya it is the third seat (the seat of virtual power) that has installed itself in the seat of formal power. At the practical level, this came about through the procedure of *self-legitimation* of the regime and its head, Kaddafi. By vehemently claiming to represent authentic Islam, Kaddafi established direct links to the seat of authority (Allāh), although he interpreted it according to his own purpose, which was to fuse Allāh with the people.

The fusion of Allāh with the people

In Kaddafi's theory, the people constitute the seat of authority. But it is not clear by what right and for what reason the occupancy of this seat reverts to the people. Is it because the people in one Muslim tradition (e.g. Khārijism) are considered the vicar of Allāh or is it because popular sovereignty is a value in itself, as in the Western democracies?

An analysis of Kaddafi's statements indicates that not exclusively one or the other, but both of these are the case. But accepting the Koran as the law of society can lead to only one result, namely, the acceptance of divine sovereignty, as opposed to popular sovereignty. How then can the conflict between these two sovereigns be resolved? There is only one way, and that is to invest (in the political domain) the people with the same attributes and the same prerogatives usually attributed to Allāh, to consider the people the vicar of Allāh, which amounts to the same thing.

Second, Kaddafi's political ideas have not been unaffected by the mainstream of European thought. For example, the political philosophy of Rousseau must have had some influence on Kaddafi, however implausible this might seem, considering his vehement and biting criticism of Western democracies. A closer reading will show that his criticisms have concerned themselves exclusively with the practice of democracy in the West, not its essence, i.e., the "machinery of government," and not its principle. According to Kaddafi it is the "substi-

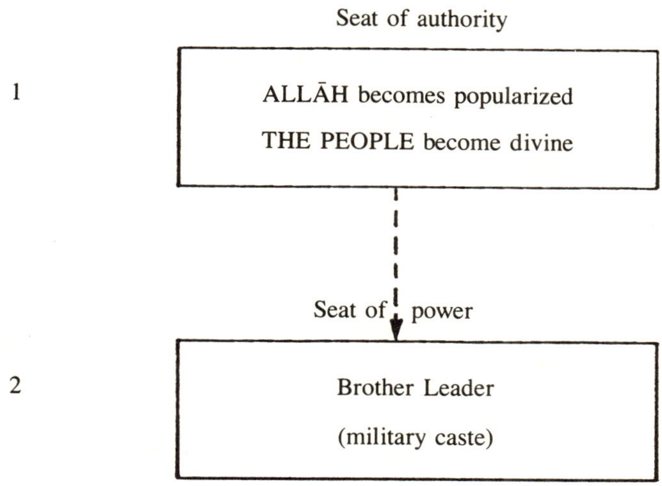

Figure 10. The Libyan Model of Power

tutes'' and ''representatives'' who vitiate democracy. Consequently, ''representation is only an imposture,'' and true democracy is one which admits of no ''substitute for the power of the people''; put differently, direct democracy yes, but national representation no. Kaddafi's rejection of national representation has two immediate consequences. At the religious level, the proclamation of direct democracy (in theory) enabled the regime to divest the *'Ulamā'* of all political and even religious representative functions. At the same time, Kaddafi used the theory of ''the people without representatives'' as a ''justification'' for the absence of parliamentarism in Libya. The parliamentary system so vigorously condemned by the Libyan leader has been replaced by a system of ''popular congresses and committees'' whose members are ''chosen'' (and not ''elected''), with the entire system under the acknowledged authority of the ''brother leader.''

The ''brother leader'' and his military caste

In Libya today, real power is concentrated in the hands of the ''Brother Leader'' (*al-Akh al-Qā'id*) in the person of Colonel Mu'ammar al-Kaddafi. This is an incontestably personal power, wielded in the name of Islam. In this respect Kaddafi's power largely resembles the power of Khomeini. While in Libya it is the ''Brother Leader'' supported by the military caste that has the monopoly on power, in today's Iran it is the "leader" and his religious caste that enjoy that monopoly. But there are also differences between these two regimes as regards effective power. In the first place, in Iran, for Khomeini, the seat of authority belongs unalterably and exclusively to Allāh. Kaddafi, on the other hand, while not contesting Allāh's authority, identifies it wih the authority of the people. The political implications of these two theoretical approaches are considerable. In Iran, the result was to give power to the ayatollahs, while in Libya power went to the colonels. The second difference is in the seat that enforces effective power. In the traditional model of Islamic power, it is the fourth sphere (the intermediate sphere) in Iran, while in Libya it is the third sphere (the sphere of virtual power) (Figure 10).

The Mujāhidīn-é Khalq model

The political theory on which the Mujāhidīn organization is based is a fusionist theory, the distinguishing characteristic of which is the fusion of God in man and *vice versa*. The choice of such a theory would be justified by the fact that the Mujāhidīn take their inspiration from diverse Muslim intellectual and political tendencies that have competed with one another over the course of history. They have likewise been influenced by a number of secular theories, nationalist and Marxist.

Inspired by different sources, their task has been both ambitious and ambiguous, namely, to come up with a synthesis of all these disparate elements.

Their effort has earned them various accusations from different quarters: from the Shah's regime, from the Khomeini regime, and even from a number of organizations currently in opposition. Thus, the Mujāhidīn have been variously accused of being "Islamic Marxists," "Khārijī," hypocrites (*munāfiqūn*), and fusionists (*iliqāṭī*).

But it is difficult to avoid seeing ambiguity in the political theory of the Mujāhidīn. This ambiguity, the result of their multiple political and intellectual inspirations, is apparent in the declarations, doctrinal programs, and political positions taken by the Mujāhidīn (alliance with Khomeini, then against him, a middle-of-the-road position between a vacillating socialism and the commercial bourgeoisie, etc.).

The same doctrinal ambiguity is likewise present in the model of government they hope to establish in Iran. In the program they published on September 27, 1981 the Mujāhidīn declared themselves for the establishment of a Democratic Islamic Republic. The contradiction in these three terms should be obvious. A *republic*, as we have shown elsewhere,[114] is not exactly *Islamic* and even less Shī'ī,[115] and *democracy* is difficult to reconcile with *theocracy*, especially in its "Islamic" form.

A close examination of the program of the provisional government of the Democratic Islamic Republic reveals many ideological currents: Islamic, Marxist, secular, nationalist, and even patriotic. Terms such as "Islam," "nation," "people," "Fatherland," and "class struggle" are used indiscriminately. In two basic themes the Mujāhidīn differ from Khomeinism: (1) the identification of divine sovereignty with popular sovereignty, and (2) the definitive elimination of the intermediate sphere (the *'Ulamā'*) as such.

Identification of divine sovereignty with popular sovereignty

The Mujāhidīn theory of power is primarily based on the occupancy of the seat of authority by God and the people (*khalq*) simultaneously.

Outwardly, this paradigm seems to resemble the Algerian system. But in the Algerian model, although God and the people both occupy the seat of authority, they each retain a sphere of autonomy. i.e., there is still a line of demarcation between God and the people within the seat of authority itself.

For the Mujāhidīn, on the other hand, this line of demarcation disappears: God and the people are fused. Thus, *politically* (and we emphasize politically), God and the people constitute one and the same entity. Nothing can separate them, just as nothing can come between them, except, of course, the Prophet and the venerated Shī'ī imāms.

The very title the Mujāhidīn have chosen for their organization, Sāzmān-é Mujāhidīn-é Khalq-é Iran, clarifies and to a certain extent substantiates this point. The terms *Mujāhidīn* and *khalq* are fraught with important implications and require some explication.

Mujāhidīn is incontestably a term from the Koran. Furthermore, the Koranic verse, a fragment of which is entwined around the coat of arms of the Mujāhidīn, was deliberately chosen to confirm the Koranic origin of the name. It reads: "Allāh hath placed the Mujāhidīn a degree above the non-Mujāhidīn . . ." (IV, 97/95)

The Mujāhidīn's choice of this specific passage is significant, for here the term Mujāhidīn is used in a neutral sense. If one considers the particular Koranic context and the verse in its entirety, there can be no doubt that the term is applied exclusively to the Mujāhidīn Fī Sabīl Illāh (Fighters on the Path of Allāh). Here is the verse in its entirety:

> Those of the believers who do not fight—except for those who have some injury—are not the equals of those who fight (Mujāhidīn) with material goods and their person on the path of Allāh; and Allāh hath placed those who fight with their material goods and their person a degree above the non-fighters. To all, Allāh hath promised a Great [Reward, but] Allāh has placed the fighters above the non-fighters in [His] immense retribution. (IV, 97/95)

The cardinal point is that the Mujāhidīn of Iran see no difference between the "*path of Allāh*" and the "*path of the people*." There can be no difference, no conflict, between these two paths. The two paths are completely identical, they intersect, merge, and become one.

While the word *Mujāhidīn* has a solely Koranic meaning, *khalq* has a dual significance: one Koranic and one Marxist (in the Iranian context). In the Koranic sense *khalq* is applied literally to the "creatures" of God. But although *khalq* is a term from the Koran, it was not introduced into Persian literature as a political expression. Besides, it is doubtful that this word is used in its political sense by men such as 'Ali Sharī'atī or Rouhollah Khomeini.

The word *khalq* was first used in a purely political sense by Marxist writers before World War II who later founded the Tudeh party. It is undoubtedly because of their usage of this word in its dual significance that the Mujāhidīn-é Khalq have been accused of being fusionists (*iltiqāṭī*).[116]

The fusionist vision of the Mujāhidīn is clearly revealed in their various declarations, and especially in their government program of September 27, 1981. In a declaration published July 2, 1981, in response to Khomeini's accusations (which called them, among other things, non-Muslim), the Mujāhidīn were solemnly obliged to renew their unshakeable faith in Islam. Thus, they proclaimed, like all revolutionary Muslims in Iran, their belief in the principles accepted by all the prophets and venerated imāms, i.e. the Oneness of God (*Tawḥīd*), the Prophecy (*Nabuwat*) and the Last Judgment (*Ma'ād*). Revealingly, these principles, so forcefully evoked in this declaration, were not taken up again in the program of the Provisional Government of the Democratic Islamic Republic of Iran, in which one can search in vain for the slightest reference to divine sovereignty or the Oneness of God, which, on the other hand, is explicitly proclaimed in the Khomeinist

Constitution in 1979 (General Principles, Article 2).

The program, however, has repeated references (thirteen direct references)[117] to "popular and national sovereignty." According to their program, the Mujāhidīn undertake to "transfer sovereignty to the people of Iran and establish the new national and popular sovereignty" (Article 1). As regards *legitimacy*, it is stated that "this government derives its entire legitimacy from the just resistance of the people of Iran against the Khomeini regime . . ." (Article 2).[118] The first temptation on reading this program is to call it a secular program, but it is not. The fundamentally "Islamic" quality of the Mujāhidīn movement in Iran must not be disregarded or underestimated. The Mujāhidīn have always claimed quite forcefully that they embody what they call "the true Muhammedan Islam" (*Islam-é Rāstīn-ē Muḥammadī*).

Actually, the real problem lies with their interpretation of Islam. Their "true Islam" is an Islam in which God and the people are merged into one, politically and definitively. Consequently, in political terms, the people are in the final analysis no less than the specific, temporal, and political incarnation of God.[119]

The Mujāhidīn have drawn on this interpretation in their attempt to resolve the problem of popular democracy and Muslim theocracy. These two different types of regimes can be reconciled in the Democratic Islamic Republic,[120] and such a republic can be realized only by eliminating the intermediate seat of power.

The elimination of the intermediate sphere

In the tradition model of "Islamic" power, the *'Ulamā'* play an important role in the transmission of effective power. The Shī'ī *'Ulamā'* can in theory play an even more important role on the strength of their claim to have the right (after the date of the occultation of the Twelfth Imām) to incorporate both the intermediate sphere and the seat of effective power. Their occupancy of the latter is considered temporary, and will end when the Hidden Imām reappears. But this is actually "a temporary measure with no end." This theory, which has now become a reality with Khomeini's accession to power, is also based on the Koran, and more particularly, on the notion of *Khalāfat Allāh* (the lieutenancy of God). The *'Ulamā'* gives a restricted and exclusive interpretation of this notion. As they would have it, Allāh, by the grace (*lutf*) that is unique unto him, has designated expressly appointed lieutenants on earth. After the decease of the Prophet, these lieutenants were the

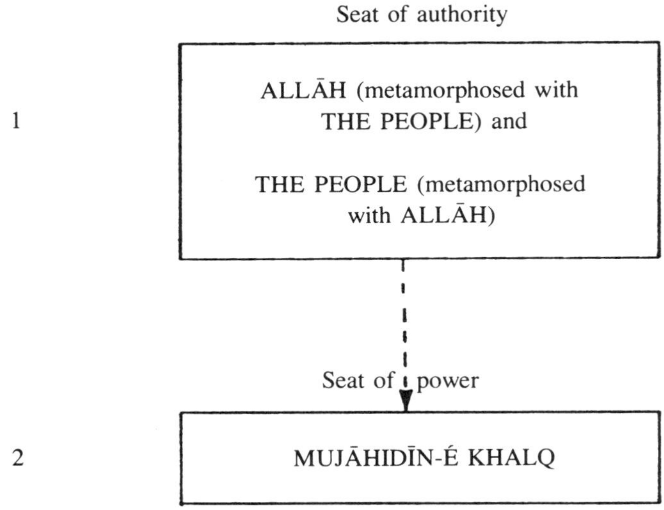

Figure 11. The Mujāhidīn-é Khalq Model of Power

twelve imāms. Since the Twelfth Imām is absent for the moment, it will thus be the *'Ulamā'* who become the lieutenants of the imām (*Nāyeb-é Imām*).

The Mujāhidīn render the notion of Khalāfat Allāh differently, giving it an extensive meaning and refusing to recognize the *'Ulamā'* as the legitimate successors of the imām. "We have never accepted, we do not accept, and we will never accept the acknowledgment, in the person of Khomeini, of any successor to the Prophet or a replacement of the Twelfth Imām . . .,"[121] they declare. If in this line of reasoning the *'Ulamā'* are neither the successors of the imām, nor even less the vicars of God, who then could indeed make a claim to be so? The Mujāhidīn have a clear answer to this question: It is the *khalq* that is the vicar of God (see Figure 11).

Thus, by eliminating the *'Ulamā'* from the sphere of power, the Mujāhidīn have in a sense joined the Khārijī and the Kaddafī.[122] The former, we may recall, did the same with the Qurayshī, who had monopolized power in the name of a *ḥadīth* deriving from the Prophet, which says that "the emirs will be from the Quraysh tribe." For their part the Khārijī have another conception of power: It is not the Quraysh tribe, but the *Umma* who have a right to effective power. And for the Mujāhidīn, the *Umma* is the *Khalq*.

Both the Khārijī and the Mujāhidīn have been accused of being *munāfiqūn* (hypocrites). It may be recalled that the *munāfiqūn* were a group of Muslims who ventured to challenge some of Muḥammad's political decisions during the Prophet's lifetime, and paid for their insubordination with their lives. The Mujāhidīn were to suffer the same fate, fourteen centuries later.

Returning to the Mujāhidīn, we might add that Khomeini has other weapons (of Shī'ī calibre) at his disposal, for example the *ḥadīth* reported by 'Alī, found in the two *Ṣaḥīh*.

> "I have heard the Prophet say: *Young men* with chimeric dreams will emerge at the end of time and proffer the best of words, but their faith will go no further than their throat. They will leave religion like the arrow leaves the bow. *Wherever you find them, kill them.* He who kills them will be rewarded on the day of judgment."[123]

4. Islam and Secularization

Toward a secularization of power

Secularization is not a recent phenomenon in Muslim societies. It has a long history going back to the first periods of Islam. The Khārijī under the fourth Caliph ʻAlī replaced in theory the caliphate by the *Umma,* which, indeed, was a major step on the way toward the secularization of the seat of power. Later, the Muʻtazilī, under the influence of Hellenism, would go even further, and extend secularization beyond the seat of power to the seat of authority itself. Thus it was that the sacred was replaced by Reason. Still later, the sacred nature of Islamic power would be challenged by the sociological approach of Ibn Khaldūn.

After the long period of intellectual stagnation attendant upon the general decline of the Muslim world, secularizing efforts would begin again in a quite new international setting, both to emulate the colonizer and to combat him. Men such as Al-Afghānī (1838–1905), Muḥammad ʻAbduh (1849–1905), ʻAlī ʻAbd al-Rāzīq (1888–1966)—the author of *Islam and the Foundations of Power* (1925), ʻAlī Sharī'atī (1933–1977), and Muʻammar al-Kaddafi multiplied the efforts.

Secularization, however, is not a linear process; its path has been erratic and its forms diverse. Each society has its own way and its own style of secularization. Recent events in the Muslim world have fueled the illusion that all secularizing efforts are doomed to failure, which is at variance with both the facts of history and with present realities. It is obviously true that Muslim societies have encountered obstacles in their secularizing efforts, and sometimes have even been forced to shift their course. Secularization has also known its discontinuities and caesurae, such as ended with the establishment of the Islamic regime in Iran, and there will probably be others in the future. But the general fact remains. Though sporadic reversals may take place at the surface, the process of secularization continues its normal course. Contemporary models, even many of those that explicitly proclaim themselves Islamic, are striking examples of this deep-seated process.

For the sake of clarity, let us distinguish between secularization from internal sources, and that which is initiated from without. By the former we mean all secularizing efforts undertaken by Muslim groups which, while remaining part of the Muslim Community, and still invoking authentic Islam as their authority, have tried variously to divest the seats of authority and power of their sacred quality. Put differently, these Muslims do not reject Islam, but endeavor to use it with a view toward establishing a non-Islamic regime, often in the name of Islam. Examples are, in addition to the traditional movements mentioned earlier (Khārijism, Mu'tazilism, and others), all the modernist and secularizing movements that have spread throughout the various Muslim countries, especially the Ottoman Empire, Egypt, and Persia, since the end of the nineteenth century. Whereas the Ottoman caliph-sultan was induced to accept the *tanẓīmāt* (reorganization) under the intense pressure of secularizing movements, Egypt for its part experienced an extraordinary intellectual effervescence under the combined effects of the reformist (*salafiyya*) and secular movements. The same occurred in Persia, where similar aspirations were to move toward a constitutional revolution. The process of secularization was completed in the Ottoman Empire when the latter was replaced by modern Turkey. Egypt and Persia were only partly successful in this regard. In the 1906 Persian Constitution, Allāh was very much present in the seat of authority, although he was no longer the sole occupant and had henceforth to share it with the Nation. This was the first of the dualist models of national essence.

The second wave of secularizing movements, still within Islam, brought new contemporary models, in particular fusionist models. Whereas the dualists undertook to separate Allāh and the Nation, the fusionists—doubtless in consideration of the unsuccessful dualistic experiment—opted for another path toward what was ultimately the same destination. It would not be misleading to say that in fusing Allāh with the nation, the fusionists, practically speaking, nationalized Allāh.

Another aspect of institutional secularization is illustrated by the efforts of men such Reza Shah (1925–1941) in Iran and Amanullah (1919–1929) and Mohammed Zahir Shah (1933–1973) in Afghanistan. But a secularization of this type is more like the Kemalist model. It would be more exact to place these last efforts at the margin of or outside the Islamic framework.

Let us proceed to an examination of some secularizing efforts of external origin. The most notable example is unquestionably that of Turkey. The Turkish experience is unique in its kind to the extent that it was a total secularization that affected power, institutions, and lifestyle all at once. Other undertakings, falling roughly into two categories, were later directly or indirectly inspired by the Kemalist model. The first comprised projects attempted by the laity, though remaining within an Islamic constitutional framework. Since any radical alteration of this framework was difficult, if not impossible, the tendency here was toward a gradual secularization of institutions (public education, implementation of a civil code, civil courts, etc.). Parallel efforts to alter lifestyle (restrictions on wearing of the *chador* and turban, liberalization of private life, etc.) were undertaken. Such on the whole was the policy pursued by Pahlavi in Iran, and to a certain extent by the Afghan kings as well.

The second group comprises projects that earlier we classified as semi-monist models. Although these models are in theory more secular than the models of Pahlavi and the Afghan kings, their actual attitudes did not always follow their secular conviction. Pragmatism generally determined their politics, which could change quite brusquely, depending on the general situation, with a flexibility that was sometimes mistaken for want of firmness. These models nevertheless remained basically secular models tending toward monist models that were national or confederate in essence. The Algerian, Tunisian, Syro-Iraqian, and to a certain extent the Nasser model belong here.

Despite differences, secularization, whether of internal or external origin, is identical on one point: It has always sought to eliminate the sacred from the system of government. The obstacles to secularization are of course numerous. Hence the slowness of secularizing movements, the ambiguity of some and the incertitude of others. Christian societies have taken whole centuries to separate the state and religion, and the process is far from complete. Surely it is asking too much, therefore, to expect Muslim society to do the same over a period of a few years or even decades, especially considering that they are still under the influence of domination and dependence.

Civil society and the nation-state

The secularization of power is a precondition for the establishment of civil society and the nation-state. A study of the contemporary models of power in Muslim society reveals that the great majority of them opted for processes of secularization as a means achieving civil society and the establishment of the nation-state. There are several reasons for this, but we shall here focus on two before we go on to discuss the obstacles encountered by these secularizing efforts.

First, the idea of a nation-state was adopted to provide a propitious general framework for ensuring the success of modernization. Since modernity originated in societies in which the nation-state was the dominant form, Muslim societies have also tended to adopt this form as a guarantee of success in their modernization policies.

Later the idea arose that the nation-state would facilitate the union, if only temporary, of the disparate forces struggling to liberate their fatherland from the colonial yoke. Thus nationalism was able to go beyond tribal, ethnic, linguistic, confessional, and social compartmentalization to give substance to a generative movement. However, in Muslim societies this nationalism had to vie with another force, which seemed to be more powerful and more entrenched than itself, namely Islam. Thus for more than a century and a half these two generative elements were simultaneously both rivals and participants. As a result, most of the progress of the Muslim national movement (the two terms "national" and "Muslim" are already inherently contradictory) has been blurred and ambiguous. Consequently, the nation-state has been affected by this original ambiguity—of which enlightened Muslims were well aware from the outset, and hence lost no time in setting about extricating nationalism from Islamism.

Theoretical efforts to this end were especially conspicuous in Egypt and Persia. In Egypt, in the *Nahḍa* movement, a theoretician of the stature of Ṭahṭāwī (1801–1873) devoted himself to acquainting his compatriots with the culture, literature, philosophy, and political system of European societies. Translations and publications proliferated. To be sure the end in mind was to arouse Egyptian society from its slumber and to open it to new ideas. But did this call for a "new culture" not also contain the implicit objective of establishing a counterweight to the traditionalist culture, religious in essence? The evidence seems to indicate that this indeed was also one of the ends envisioned. In their theories, Ṭahṭāwī and his disciples drew a basic distinction between the notion of *Umma*, Islamic at root, and the notion of *waṭan*, literally, fatherland. The word *sha'b* (nation) had not yet come into use[124] since this was only the first stage of a national awakening, that of acquiring "a self-centered autonomy within an empire of purely formal ties."[125] In other words, a *territorial* autonomy (the fatherland), with full-fledged independence as the ultimate goal. The use of *waṭan* in place of *sha'b* entailed other advantages as well. The fatherland (the soil) took the lead over Islam (the faith). Indeed, those who proposed to give priority to Islam found themselves utterly disarmed by the use of a term as neutral as fatherland, which, in fact, the Prophet himself had bid them to love as a *ḥadīth* (*ḥūbb ul-waṭan min al-imān*).

The theoretical disengagement of nationalism had been preceded by the attempts of Muḥammad 'Alī to establish an economy of national scope in place of the local economy. His works clearly espouse the future construction of a nation-state. However, such efforts remained unsuccessful and the young Egyptian economy was very quickly brought back into the fold and absorbed into the system of world domination. Anwar 'Abdal Malek sums up this process of integration as follows:

> Thanks to Muḥammad 'Alī, Egypt was given a national and not merely a local economy, an economy that was both state-controlled and autarkic. The great works that had been accomplished under Isma'īl structured the internal market. It was loans and cotton as a single-crop culture, initiated and deepened by colonial penetration and imperialist occupation, that achieved Egypt's integration into the world market over the course of the last third of the nineteenth century.
>
> The distortions imposed on the Egyptian economy were severe and with lasting consequences: incorporation into the world market under European hegemony on terms imposed by imperalism; the essential dismantling of state monopolies in industry; the forced link to sterling; horizontal division of the country between Lower Egypt, open to modern technology, and Upper Egypt, which remained relatively apart, lines of transport and communication having the sole end of accelerating the routing of harvests to Mediterranean ports.[126]

During this same period, Persia underwent an evolution almost identical to that of Egypt. Although unlike Egypt it enjoyed a formal independence, it too was hardly immune to the effects of domination and the interventions of foreign powers in its internal affairs. Again as in

Egypt, a commercial bourgeoisie began to develop in the second half of the nineteenth century and endeavored to cement the foundations of a new national economy in its own interests.[127] Thus it was that the various Iranian trading companies were created, and the land, which traditionally had belonged to the King, began to be invested in as private property. The commercial bourgeoisie profited from the sale of lands and public domains. It began to organize itself. A Chamber of Commerce was established, first at Tehran (in the 1880s) and then in the major cities and the provinces. With the rise of the commercial bourgeoisie—and indeed to a large extent because of it—rebellions began to take place against the economic concessions (tobacco, railroads, oil, communications) accorded to foreigners, ultimately to end in the constitutionalist revolution of 1906. With the establishment of a constitutional regime, a state, tribal in essence (the Qājārs), was replaced—at least in theory—by the nation-state.

Parallel to the development of economic autonomy, the Iranians, following the same course as the Egyptians, attempted to steer their country toward political modernization. The word *mellat* (nation) began to appear in Iranian political literature for the first time, counterposed to the despotic prince, but also to Islam. Reality of course was more complicated. For example, in its conflict with the authoritarian regime of the Shah, *mellat*,[128] at the time represented by the commercial bourgeoisie, was driven into an alliance with a number of the *'Ulamā'* who were also in opposition, although for other reasons (financial reasons as well as reasons of influence and prestige). The ultimate consequence of this tactical alliance was a fundamental confusion, of which the 1906 constitution, with its dual source of legitimacy (Allāh and *mellat*), is elegant written testimony. But that was not to be the end of the matter: The confusion had grave consequences later on as well for the political life of the country. The fall of Mosaddeq (nationalist—*melli*—leader) in 1953 was due in part to the conflict that opposed him to his former ally, Ayatollah Kāshanī (religious—*madhhabi*—leader and mentor of Ayatollah Khomeini).

The fall of the Mosaddeq government was followed by an imperial regime that lasted twenty-six years, i.e. down to the fall of Pahlavi. The project of the nation-state set forth in the 1906 Constitution was discredited totally, thereby leaving a broad opening for the spread of religious discourse. The commercial bourgeoisie (*bazar*) tended toward traditionalist sources (the *'Ulamā'*) with a view to blocking the rise of the new oil bourgeoisie, which was beyond the pale of influence of the *bazar*. Iranian political life became polarized between the supporters of the imperial regime and the supporters of an Islamic regime. Once established, the latter would pursue a policy toward the partisans of the nation-state that was far more severe than that of the Shah's regime. Nationalists were persecuted, imprisoned, and executed.

Is Islam an obstacle to the building of a nation-state?

Having reviewed the reasons for the option of a civil society and a nation-state and the various obstacles that this option encountered, let us now turn to the fundamental question of Islam's relationship to civil society and the nation-state, so as to determine at what level and to what extent Islam is in contradiction with them. This question has occasioned impassioned debates both within Muslim society and outside it. Indeed, there are many who believe that Islam, by virtue of its singularity, is the antipode of civil society and the nation-state. This "singularity thesis" has the weakness that it regards Islam as a unified bloc, a completed system, an indivisible totality. Clearly such an approach, which is basically culturalist, can only produce sweeping final judgments about Islam and Muslim societies. The conclusion was thus reached that, because of the presence of Islam, Muslim societies could never build a civil society. Such a conclusion is possible only if one wholly disregards the impact of external factors on Muslim societies. Moreover, the anti-evolutionary bias of this approach impels it to set up the "law of the *status quo*" as an eternal law.[129]

A more plausible and realistic approach would regard Islam as a human construction and, like all other such constructions, subject to change and evolution. Islam too will undergo changes and modification if the conditions are ripe for them. The Islam of today is not the Islam of yesterday, and the Islam of tomorrow will probably be even more different. The existence of a legion of Islams today, each of which holds itself to be the one and only authentic Islam, is sufficient evidence that Islam is neither a unified bloc, nor an edifice that will weather all eternity without change. To arrive at some understanding, however incomplete, of Islam's relationship to civil society and the nation-state, it is first necessary to draw a distinction between the conception and the *reality* of Islam. At the conceptual level Islam does indeed contain a number of notions that could be construed as obstacles to the establishment of civil society. In the Koranic system, for example, there is no place for the *status of citizen*. The Community is made up of believers, not citizens. Another example is the symbiosis between the temporal and the spiritual. It matters little if the symbiosis is genuinely of Koranic origin, or if it has been fashioned over the course of history. The essential point is that, in Muslim consciousness, there is no distinction between Allāh and the prince in Islam. Furthermore, this idea can influence people's political choices. The notion of Allāh itself can serve as a barrier to the construction of civil society. As history will abundantly testify, caliphs, kings, emirs, shaykhs, and ayatollahs have always been able to invoke Allāh as a pretext for refusing to recognize the civil rights of individuals.[130]

These and similar examples demonstrate that Islam does indeed have concepts capable of obstructing the evolution of Muslim society toward civil society. But it would be mistaken to think that these concepts have in themselves been determining factors in the failure to achieve the civil society envisaged. The establishment of the Islamic regime in Iran has often been cited as the example *par excellence* of the conceptual powers of Islam. Since this example is currently the trump card of the "culturalists" for validating their own arguments, it may be useful to dwell on this aspect of the Islamic revolution in Iran. My purpose here is to demonstrate that, contrary to what is generally thought, Islam at its conceptual level, or as an ideology, has not played the role mistakenly attributed to it by some in this revolution. I shall limit myself to the most concrete and verifiable examples.

Those who believe in the autonomous powers of Islamic concepts overlook (or pretend to overlook) two things. First, as Amilcar Cabral put it, "people do not struggle for ideas or abstract notions; people struggle and accept sacrifices to obtain material advantages so as to be able to live in peace and prosperity. People want to see their lives progress and the future of their children assured."[131] Finally, referring to a specifically Iranian context, I wish to stress that Islam is not a religion that has recently been introduced in Iran. The country has been Muslim for fourteen centuries. It has been officially Shī'ī since the Ṣafawid (1570–1747). Consequently, if Islamic concepts were efficacious in their own right, Persians would have been living for a long time already under a regime comparable to that of Khomeini.

Even if we assume that, for one reason or an other, these concepts had for fourteen centuries been unable to act and acquire concrete form in an automonous Islamic political regime, and that they first acquired this possibility quite recently, the study of the remote, recent, and immediate origins of the "Islamic" revolution shows that the ideology of Islam was scarcely one of its fundamental causes. However, it not being my purpose here to analyze the revolution, let me merely enumerate some of the real causes of it. I shall concentrate on internal causes, leaving the external ones aside for the moment.

Among the deep-lying causes there was the fundamental question of the *system of property*. Traditionally, land, water, and pasture, the three essential elements of the economy, belonged in the main to the prince.[132] Privatization of the land goes back no further than the nineteenth century. This change played a major role in the formation of the new class that would later be the driving force of the constitutionalist revolution, but it would also give rise to a social ambiguity. The new class was not an autonomous class. In Western terminology it would be somewhere between a feudal class and the bourgeoisie. But with the appearance of oil as a new and essential element in the national economy, the situation reverted very quickly to that of the preconstitutional period. With the fall of the Mosaddeq government, the nationalized oil was effectively restored to the international cartels. Through the "second nationalization" under the Shah's regime, followed by a substantial rise in the price of oil, the Shah, strengthened by his new oil resources, was, like the preconstitutional Iranian kings, able to intervene in all spheres of social, economic, and political life. Of course this was in clear contradiction to the Constitution. In 1973, the Shah, Chief of State, became the Leader, and above all the great Employer. Thus the path to civil society was blocked once again, this time, however, not by Islamic concepts but by an imperial regime.

As for the more immediate causes of the revolution, there were quite a number of them, but one would seem to be the most important: the disparity between economic development and political development. While oil had put some considerable financial means in the hands of various social groups, the Shah's authoritarianism not only did nothing to facilitate the political participation of these groups, but he made every effort to deprive them of all means of political expression. The inevitable result was a broad disjunction between economic and political life. Hence the "unexpected" and widespread discontent in all social strata, even the most privileged.

Disregarding various other factors (corruption, overcentralization, a single-party system, arbitrary imprisonment, torture, foreign intervention, etc.), we come finally to the immediate causes, likewise manifold. One of these was an electrical power outage. To include such a factor, at the surface more technical than political, among the causes of a revolution so vast in scope might seem fanciful, if not misplaced. How, it may be asked, could an electrical power failure contribute so amply to the triggering of a revolution so original and so immense that its advent astonished the entire world? The answer is simple enough: because the power failure in the summer of 1977 not only spread discontent even into the privileged strata, who were particularly outraged (as they were much more dependent on electricity than the less-privileged strata), but also because this failure brought to light the weakness, incoherence, and vulnerability of a regime until then reputed to be a powerful and cohesive force.

Let us take a closer look. The Shah boasted of having built a mass consumption society. This was true. The Iranians had never consumed as many things and in such a variety as during the last ten years of the Shah's regime. They consumed without thinking and spent without counting. All went well until the system was plunged into serious economic and financial difficulties, in the midst of which occurred the long power failure in the summer of 1977. For reasons as yet unknown, the central power station constructed by the Shah's regime in conformity with his policy of centralization began having troubles, and consumers suddenly found themselves with electricity for only two to three hours per day, and very erratically at that. With temperatures between 35°c and 50°c, life became unbearable, particularly for those who had become used to electrically powered equipment and home appliances. Fans, elevators, refrigerators, televisions, and similar items on which the Shah had built his regime suddenly became superfluous and their mere presence in homes and public places was enough to inflame the tempers of anonymous consumers to the boiling point. Meat, imported at great cost from Australia and New Zealand, spoiled as the situation persisted. No freedom and no consumption—that makes for rebellions if not revolutions, especially when a suffocating heat stokes spirits. Everyone, from the Shah down, began looking for scapegoats. A committee of imperial inspection was even created to this end. Heads rolled (dismissal and then imprisonment of the Prime Minister, other ministers, generals, businessmen), but this hardly improved the situation. Then it was the turn of the Shah's own family; his relatives discreetly left the country one by one. No progress. Finally, and ineluctably, as has happened so many times before in history, it was the turn of the Shah himself. His head too fell as a symbol of all the ills and misfortunes of the thirty-seven years of his reign.

I would of course not go so far as to say that the 1979 revolution was caused by a power failure, but this example does demonstrate how an outwardly banal event can lay bare an extremely powerful and cruel regime, and even bring it to its knees. It should be clear that the power failure had a much greater impact in the first phases of the revolutionary upsurge than any religious discourse. It is equally clear that the Iranians did not have a revolution because they were unable to fully perform their religious duties (which as a matter of fact they were able to do undisturbed) or even because the Shah's regime did not have the characteristics of an Islamic regime. The deep-seated causes of the revolution must be sought elsewhere, as we observed earlier. In the situation in which the Shah found himself in 1978, the regime was doomed in any case, with or without Islam, with or without Khomeini or some other popular leader.[133]

Why then, we may ask, did a rhetoric couched in religious terms, rather than some other, gain the upper hand? There are two fundamental reasons for this. First, the Iranians, weary of the Shah's regime, had their minds set much more on its departure from the scene than on the nature of the regime that should follow it. No political group, and no leader, had worked out a program for change. Actions were ill-defined and equivocal. Khomeini alone had a blueprint for government and a plan of action. While others based their actions on conjecture and the evolution of events, Khomeini had a program he had worked out beforehand, and he followed it. He won out quite simply because he knew where he was going. There is nothing mysterious about this; one need only restudy the positions adopted by the Iranian political groups and leaders from 1977–1978 onward and compare them with Khomeini's. It is incontestable that Khomeini was able to give an historical dimension to a specific political struggle. Moreover, he managed to reconcile the different social classes, if not permanently, at least long enough to forge an alliance among them against the regime in power. In addition, he showed himself to be courageous both physically (he arrived in Tehran in a situation that was highly unstable) and morally. These are the characteristics Machiavelli called the virtues of a leader.

The second reason the religious rhetoric prevailed over the others (socialist, nationalist, modernist) was because it managed to become the most impelling rhetoric within the overall context of the revolutionary movement. Thus, the great majority of Iranians believed at some given moment that the various other subrhetorics could be integrated without clash or conflict into the great religious rhetoric. In the general confusion, each tendency couched its own rhetoric in the same terms as the one being propagated in the name of Islam. It was at this time that the rumor began to circulate that Khomeini's visage could be seen in the moon, and people climbed to the rooftops to see it with their own eyes. Thus the Iranians, abandoning their areligious rhetoric to oblivion, in the belief that they had finally found the one universal Rhetoric, found themselves suddenly on the moon. Each construed Khomeini's rhetoric as he wished, and as we know, it did indeed contain something for everyone, at least for the moment.

Conclusion

In each society one finds competing persons, groups, and classes that already hold effective power or are struggling to gain it. In the end it is these rival interests that decide the form, the character, and the magnitude of the "Islamic ingredient" in power.

The unending appeals to universal Islam have no ultimate meaning other than that given them by the conquest and maintenance of power. Further, since each specific form of Islam (born of the particularity of each Islamic regime) cannot have the legitimacy necessary for consolidation of power, "Islamic" regimes must go beyond the confines of their own specific Islam and perform as authorized and exclusive guarantors of universal Islam.

Thus, every specific, limited, and dated form of Islam, every Islam made to measure, readily calls itself simply Islam, i.e., universal Islam, thereby implicitly denying that quality to each of the others.

These unending appeals to universal Islam, shrouded with a deceptive innocence, conceal the true purely temporal intentions of those actual or potential holders of "Islamic" power.

For example, we know how vehemently Ayatollah Khomeini insists on the univeralist, authentic, and true character of his Islam, and we all know the zeal with which he attacks all other models, which he is quick to call "falsifiers of Islam."

In his assumption of sole proprietary rights to universal Islam and his exclusion of the Islam upheld by others, Khomeini's appeal has been able to mobilize the Muslim masses. Millions of Muslims are prepared to sacrifice their lives for the cause and the victory of Khomeini's Islam; but this presumably universal Islam nevertheless remains just one more form of Islam, by Khomeini and for Khomeini. The statements and actions of the Ayatollah leave no doubt in this respect: Islam cannot exist without the "*faqīh* government," i.e. without the power of the *'Ulamā'*, of which he is the head. The Ayatollah has never concealed this point. He has said expressly that: "Islam without *akhond* (a Muslim theologian) is not Islam."

Similarly, "Islamic government without the control of the *Faqīh* is not conceivable" (*Kashf ul-Asrār*, p. 281). It goes without saying that the victims of such manipulation and mystification are in the first instance Muslims who believe in this universal Islam. While they are enduring immense sacrifices for the cause of Islam, the Ayatollah moves not a finger to sacrifice his *'Ulamā'* for the same cause.

What Khomeini does not do, Kaddafi and the Mujāhidīn-é Khalq do with alacrity. We have seen how Kaddafi and the Mujāhidīn have flatly eliminated from their model of power the sphere traditionally reserved for the *'Ulamā'*.

The Libyan colonel has openly challenged the foundations of the *fiqh* and the *ḥadīth*, the two pillars on which *'Ulamā'* power has been based historically. The Mujāhidīn in turn have adopted an identical attitude toward the *'Ulamā'* since their disillusionment with the Khomeini regime. Can we thus say that Kaddafi and the Mujāhidīn were guided by religious motives in their "anticlerical" bearing? Can we say that in adopting this attitude they had wanted to "rectify" the deviations produced in the sacrosanct model of Islam? And if these things are true, in the interests of what group, of what party?

Whatever the answer to these questions, one thing is certain: both Kaddafi and the Mujāhidīn are decidely hostile to the power of the *'Ulamā'*. Like Khomeini, they believe that they are invested (by whom?) with a holy mission to purify Islam of its "usurpers," "falsifiers," and "forgers," which means that they must "liberate" Islam from the grip of the *'Ulamā'*. This is the official discourse, i.e. what is said at a superficial level. Under the surface, however, the matter appears in a different light: Islam yields its place to *power*. At this level Islam becomes simply an instrument, a cosmetic, concealing the true ambitions of the competing actors. At this level the hostility of Kaddafi and the Mujāhidīn is no longer religiously motivated, but essentially a political attitude and behavior. Kaddafi and the Mujāhidīn are hostile to the *'Ulamā'* for the simple reason that they do not belong to the religious caste with which they are competing. Further, the Libyan military caste, which holds a monopoly on power, is clearly unable to come to an understanding with the religious caste, which has the same ambition. Conversely, the Iranian religious caste is not about to yield power to the Mujāhidīn (its Islamic rival) and even less to other groups, feeling quite capable of maintaining its monopoly of power on its own.

The real stake in these contests is not Islam, but power. The battle centers on the seat of power.

But this *de facto* political struggle in the name of Islam must still be justified by the medium of "Islam," and it is here that the seat of authority becomes an issue.

As a religious caste, the *'Ulamā'* have every interest in appealing to the authority of Allāh, whose will (*mashiyya*) can be correctly interpreted only by the *'Ulamā'*. In this model, therefore, there is no place either for Kaddafi or the Mujāhidīn, who, despite their appeals to Islam appeals, seek to eliminate the *'Ulamā'* from the seat of power. They are obliged to present Allāh in such a light that the claim of the *'Ulamā'* to be interpreters of the divine will becomes impossible. To do this they have fused Allāh with the people and presented themselves as the faithful interpreters and incarnation of both the divine and the popular will.

The divergent positions of the monists and fusionists is of course not without practical implications. A monist regime, religious in essence, is also monopolist and totalitarian, which makes it totally incapable of integration. Put in terms of game theory, a monist regime plays a zero-sum game, both in its internal conduct (*vis-à-vis* the various sociopolitical groups) and in its external conduct (e.g., the rigid and Armageddonist attitude of Khomeini in the Iran–Iraq war). Following the same line of thinking, the existence of a monist regime is totally incompatible with the existence of rivals. In a monist regime the notion of a social or political "partner" and, hence, dialogue, is practically nonexistent. Another characteristic of the monist regime lies in its Manichaean vision—there are only black and white: *dār al-Islam* and *dār al-Ḥarb*, *Ḥizb Allāh* and *Ḥizb al-Shayṭān*. Political behavior is governed by the principle of *tawallā* (rapprochement, friendship) and *tabarrā* (distance, hostility). These latter principles figures explicitly in the preamble of the Constitution of the Islamic Republic of Iran.

All together, these characteristics account for the arrogance (often interpreted as "irrationality") and isolation of the monist regime.

The fusionist regime is less rigid and less dogmatic than the monist regime. It is more open to compromise and cooperation. This relative flexibility and limited openness of the fusionist regime derives naturally from the plurality of elements (at least two) that make up the seat of authority. Arabism in Libya and Marxist socialism among the Mujāhidīn set the tone and charge the atmosphere for their respective forms of Islam.

At the same time, since fusionism is by nature amalgamist, the politics it engenders are uncertain and hesitant. Fusionist politics are generally extremist politics: there is vacillation between two extremes, alliance with and hostility toward the same party (as between the Mujāhidīn and Khomeini, and between Kaddafi and and Sadat).

There is one final question concerning an apparently paradoxical aspect of relations between the fusionists and the monists. If it is true that they are in a fundamental state of reciprocal hostility, how can one explain the cordial understanding that currently reigns between the monist Khomeini and the fusionist Kaddafi? There exists no political, military, or ideological alliance between these two men. Their understanding is really a tactical one, and it is aided by the considerable geographical distance separating Libya from Iran (which makes impossible any border conflict and complicates any physical intervention). Geography is thus the first tangible reason for this "entente," but other motives—such as the Palestinian problem and the politics of oil—must also be included. The two countries have effectively adopted a similar attitude toward such non-Islamic problems. Despite their tactical understanding, the Ayatollah and the colonel clearly mistrust one another. Khomeini has by no means forgotten the obscure affair of the disappearance in Libya of the Imām Ṣadr, his close relative and the politico-spiritual leader of the Lebanese Shī'ī. Kaddafi in turn has not forgotten that his prime minister Jalloud was for this reason kept "prisoner" in Iran for several days by the Ayatollah. Moreover, recent events have shown that the two men are in fact in conflict, but through various intermediaries. While Kaddafi uses every possible means to repress Libyan fundamentalists, who are the natural allies of Khomeini, especially after the failed conspiracy against his regime (in May 1984), Khomeini was not upset to see the Libyan embassy at Beirut blown up by his Shī'ī partisans in July 1984.

To sum up, it appears increasingly unlikely that the monists and fusionists can come to an agreement on developing an integrative model of Islamic government. It also appears that the Islamic "renewal" occasioned by Khomeinism and Kaddafism is more illusion than reality.

Dualist models may be divided into two categories: one in which Islam is paired with the *nation*, and a second in which Islam is paired with the *tribe*. The second, of which Saudi Arabia is the best example, has traditionally proven most resistant to shocks and to sociopolitical unrest. On the other hand, the dualist models of the Iranian (under the Shah) and Egyptian (under Sadat) stamp have seen regular and grave turmoil: the first was overthrown, and the second lost its President at the hands of assassins in the midst of a plethora of social and political woes. However, whether in times of stability or times of travail, it is difficult with regard to dualist regimes to determine what is a consequence of the established model and what is due to socioeconomic factors.

It is certain enough that the relative stability of Saudi Arabia is not solely the consequence of its Islamic and tribal specificity. Similarly, the instability of the dualist regimes of the Shah and of Sadat was not necessarily due to their dual national and Islamic nature.

Other factors most certainly are at play: demographic and geographic factors, the system of income distribution, or the state of society in general.

Whereas demography and geography play a rather stabilizing role in Saudi Arabia, with its total population of 7 million (significantly less than the population of the city of Cairo), the same factors could easily have a destabilizing effect in Egypt and Iran where a large part of the population lives in crowded urban centers. The same applies to income distribution, which is less unjust in Saudi Arabia than in the other countries. All this is further compounded by the fact that Saudi society, unlike Egyptian and Iranian societies, has not yet arrived at the stage of the nation-state and its institutions: e.g. parliamentarism, centralism, bureaucracy, etc. Indeed, Saudi society is for the moment free of this sort of epoch-making changes. Evidently, a dualist model alone, whatever particular form it may take, is not in itself a determining element. Consequently, dualist regimes are left with nothing else than the instruments of power, which are exploited by those who hold power as well as by their political adversaries at one and the same time. For example, in a tribally dominated society, it is normal for power to rely more on tribalism than on the national factor. In contrast, in countries where national sentiments are present, but not sufficiently entrenched, those holding the reins of power attempt to make use of Islam and the nation concurrently and in a contradictory fashion. This of course does not mean that the opposition does not do the same thing. Indeed, in the three cases mentioned, the various oppositions have all had the common character of attacking, for a variety of reasons, the ambiguous nature of their respective regimes. For some these regimes are hypocritical because they make use of Islam for political ends without being truly Islamic, while for others these regimes are reactionary by the mere fact that they have anything to do with Islam at all. It may therefore be concluded that dualist regimes are more outwardly vulnerable than monist and fusionist regimes and that this particularly applies to nationalist-Islamic regimes, more so even than to tribal Islamic regimes. Semi-monist models may also be divided into two groups, one represented by the Algerian model, which tends to integrate Islam into the system, without, however, according it autonomy (this might be called an integrative model) and a second, represented by the Ba'th model, in power in Syria and Iraq. This model is disintegrative insofar as Islam is *presumed* to be independent of the state.

Judging from the events of the recent period the Algerian model seems to be more resistant to internal turmoil than the model operating in Iraq and Syria. It may reasonably be asked whether Algerian stability is the result of the regime's integrative nature or whether it is due to factors of a totally different kind. No answer can be more than partial, but one thing is certain, and that is that in the Algerian revolution the *national* element was a powerful mobilizing factor, much more so than the *Islamic* element. And the situation has not changed much since independence. The *nation* continues to occupy center stage, while *Islam* provides it with a historical space and a cultural dimension, at the same time defining the contours of a collective Algerian identity.

The situation is somewhat different in Syria and Iraq, where Ba'thism, still in power, calls for a secularization of the whole of Arab society—which, put concretely, means a repudiation of Islam as a religious ideology and its acceptance solely as a historical phenomenon in the Arab cultural heritage. Hence the exclusion of Islam from the state without another potent ideology (nationalism, Arabism, socialism) being offered in exchange has thrust Ba'thism, even as it holds power, into direct confrontation with Islamic movements.

Finally, it is relevant to know whether an entente between areligious forces comprising nationalists, socialists, and liberals, on the one hand, and Islamic forces, on the other, is at all possible in contemporary Muslim societies. Obviously, given the singularities of each of these societies, no statement in this regard could be more than an educated conjecture; still, if one considers the events of the past two centuries in the general evolution of Muslim societies, it is impossible not to be impressed by the utter failure that has been the lot of any attempt to forge a lasting alliance between the above two blocs. Either such attempts have been defeated out of hand, or they have led to open conflict between the two opposing blocs. The experience of the alliance fashioned in Iran in the early 1950s between Mosaddeq and Ayatollah Kāshānī is a case in point, as is the accord that reigned during the first years of Nasser's time in power between him and the Muslim Brotherhood. Both attempts ended tragically: Mosaddeq was overthrow by a *coup d'état* following a conflict that had him squared off against Kāshānī, while Nasser (having escaped an assassination attempt prepared and executed by the Muslim Brotherhood; Sayyed Qutb, head of the Muslim Brotherhood, was arrested and executed following this incident), died, with power passing to his direct successor Sadat, who himself later fell under the blows of followers of the spiritual disciples of this same Qutb.

Historical experience also shows that in cases in which the two above-mentioned forces merged to form a common front, the end result was either a total failure of the project or a turn of events against the areligious forces. The difficulties being experienced by the Egyptian neo-Wafd as a consequence of its coalition with the Muslim Brotherhood are concrete and current testimony to this. Similarly, all areligious political personnages who entered into an alliance with Ayatollah Khomeni in the hope of giving birth to a national-Islamic or social-Islamic common front were deeply deceived by the turn that events ultimately took in the first years of the Islamic revolution.

Consequently it would appear that any alliance between areligious forces and Islamic forces is in the end unworkable. It is only in exceptional moments, such as a war of independence or national liberation or a war against a foreign aggressor, that areligious and Islamic forces have succeeded, under the pressure of circumstances, in working together toward the same end. But once that end was achieved, whether wholly or only partially, the ill-starred alliance has come apart at the seams, the parties to it ultimately passing into conflict.

On the basis of the foregoing, it clearly appears that the *concept* of Islam does not have the autonomy to obstruct the evolution of the structures of Muslim society. Yet on a real level—that is, the aggregate of the social structures and institutions erected in its name—Islam is indeed quite able to erect obstacles to this evolution. However, the structures of a society are the products of society itself. There is no evidence that Islam *as faith* is capable of producing these structures in society's stead.

Any real debate, therefore, must center on the *realities* of Islam, i.e. the basic structures of Muslim societies, and their possibilities for change and improvement. There have been many efforts in this regard, and if these efforts have met with obstacles, it has not always been because of Islam's presence. In all third-world societies, whether Muslim or not, the same phenomenon can be noted: the external factor bears much of the responsibility for their backwardness.

Every time Muslim societies have attempted to free themselves from domination, every time they have begun to build an autonomous national economy and to take control of their own natural riches, and every time they have attempted to initiate a process of democratization with the end of establishing a Western-style political regime, external forces have lost no time in mobilizing every means to ensure that these efforts would fail. Examples abound in both Muslim and non-Muslim societies—the military *coups d'ètat* against the democratic governments of Mosaddeq in Iran and Allende in Chile, to name but two. Were these governments overthrown and replaced by dictatorial regimes because traditional structures could not accommodate democratization, or was it simply by order of the successors of President Wilson, who has the honor of having invented the noble phrase "the right of nations to self-determination"? It seems gratuitous to place the blame for the failure of the grand schemes of democratization solely on the societies themselves and the structures inherent to them. To say that these structures are resistant to all change is to belabor the obvious; but the real problem lies elsewhere, in the fact that efforts undertaken in Muslim societies to establish a nation-state, a civil society, and a democratic regime have almost invariably been interrupted by the intrusions of foreign powers.

Indeed it is precisely because of these repeated interruptions that Islamism is today being reborn from its own ashes as a new alternative. And ironically, some Western theoreticians and intellectuals who consider themselves leftist see in this so-called Islamic "renaissance" proof of the appropriateness of their Europocentric theories. Making an abstraction of the decisive impact that domination has had on Muslim societies, these idea-mongers point to the structural incapacity of these societies to adopt a democratic model and found a nation-state. They ignore the fact that Christian societies, from which they themselves have descended, took centuries to arrive at the point at which they now find themselves. As far as Muslim societies are concerned, despite an unfavorable conjuncture and innumerable obstacles, they have already taken significant steps toward their objectives of democracy, secularization, and progress. However, there is no guarantee that they will succeed in this task. Success will depend in large part on the determination of their people and the perspicacity and unity of their leaders.

Notes

1. This basic distinction is argued for in particular by M. Arkoun throughout his critical works.

2. The four approaches are clearly distinct from the four fundamental Sunnī schools: Ḥanbalism, Mālikism, Shāfiʿism, and Ḥanafism.

3. Al-Fārābī, *Kitāb Āra Ahl al-Madīnat al-Fāḍilat*, French trans. *Idées des habitants de la Cité vertueuse*, p. 89.

4. Ibid.

5. Ibid., *Introduction*, pp. 8–9.

6. Ibid., p. 95.

7. For instance, we may cite the work of the Grand Wazir Niẓām al-Mulk (Ṭūsī) (508–485/1017–1092) entitled *Siyar al-mulūk/Siyāsat nāmeh*. However, Niẓām's works are centered on the question of *sultanate* (royalty) and not the caliphate. Similarly, Ibn Taymiyya (727/1328), in his work *Siyāsa shar'iyya*, takes up questions relative to administration; but this work, like all the writings of Ibn Taymiyya, is of a polemical nature.

8. H. Laoust, *La politique de Ġazālī*, p. 19.

9. Ibid., p. 234.

10. In our times, we can observe the emergence of a similar movement among a certain number of Muslims, the "fundamentalists" and the "integrists." Whether through intellectual laxity or ignorance, this movement has been interpreted as an "Islamic renaissance," especially outside the Muslim world.

11. *Iḥyā*, II, pp. 299–300, translated by H. Laoust in *La politique de Ġazālī*, p. 131.

12. Ibid., p. 369.

13. Ibid., p. 373.

14. Ibn Khaldūn, *Muqaddima*. French trans. *Discours sur l'Histoire universelle* vol. I, p. 69.

15. This is the thesis put forth by the new "sociology of the state." Ibn Khaldūn understood it six centuries ago.

16. *Muqaddima*, I, p. 53.

17. M. Arkoun, in *Islam, hier-demain*, p. 151.

18. G. Delanoue, "Une épitre de Ḥasan al-Bannā aux Frères Musulmans," p. 55.

19. George Makdisi, "Authority in the Islamic Community."

20. Laoust, *Les fondements de l'autorité dans le "Minhāj" d'al-Ḥillī*, p. 175.

21. B. Badie, *Culture et Politique*, pp. 83–84.

22. On the Constitution of the Medina see: R. B. Serjeant, "The Sunnah Jāmi'ah, Pacts with the Yathrib Jews, and the Taḥrim of Yathrib . . . ." This section on the Medinan Umma has been largely inspired by Serjeant's outstanding work.

23. Abū 'l-Faraj al-Iṣfahanī, *Kitāb al-aghānī* 19, p. 94.

24. A. M. A. Makarti, *Water Rights and Irrigation Practices in Lahij*, p. 122.

25. Arkoun, *Pouvoir et Verité*, p. 160. The Council of Ten was essentially equivalent to the Politburo in the Soviet Union and the National Security Council in the United States, allowing of course for differences of scale.

26. The use of the term *khalāfat*/caliphate in Islam dates from the election of Abū Bakr to the presidency of the *Umma*, which occurred just after the death of Muḥammad. Later the term *khalīfa*/caliph was sufficient to designate the one who held supreme power. Muslim theologians have tried to give a philosophical and theosophical meaning to the concept of caliphate.

Citing certain Koranic verses, they consider the human race in general and the prophet in particular as vicars of Allāh on earth. Let us quote two verses that deal directly with this subject: "Oh David, we have made you vicar on earth. Arbiter among in the way of truth . . ." (S. XXXVIII, V. 26) and "When your lord saith to the angels: I am going to place a vicar on earth . . ." (S. II, V. 30). A more precise idea of Sunnī theoreticians' definition of the caliphate is given by Al-Māwardī, who defines the caliphate as "the authority invested in the person who replaces the prophet in the dual mission of defending the faith and governing this world" (*Al-Aḥkām al-sulṭāniyya*, French trans. by Léon Ostrorog, vol. 1, p. 95). Ibn Khaldūn gives a similar definition: "The institution of the caliphate serves as a vicariate to the Legislator (Muḥammad) insofar as it, like Muḥammad, maintains the faith and governs the world" (*Muqaddima*, French trans. *Discours sur l'Histoire universelle*, vol. 1, p. 379).

27. Al-Baghdādī, *Kitāb uṣūl al-dīn*.

28. Ibid.

29. Ibn Taymiyya, *Siyāsa shar'iyya*, p. 172.

30. Ibid., p. 173.

31. Al-Māwardī, *Aḥkām al-sulṭāniyya*. The quotation is from a translation by H. Laoust.

32. Ibn Khaldūn, *Muqaddima*, vol. 1, p. 614.

33. Ibid., p. 310.

34. Al-Māwardī, *Aḥkam al-sulṭāniyya*, French trans. *Les statuts gouvernementaux*, p. 10.

35. Ibn Khaldūn, *Muqaddima*, vol. 1, p. 414.

36. Ibid., p. 415.

37. The members of the conclave, aside from the president, were 'Alī, 'Uthmān, Sa'd b. Waqqāṣ, Zubayr, and Ṭalḥa.

38. It is interesting that fourteen centuries later, a pretender to the position of 'Alī in the person of the Ayatollah Khomeini would come to power in similar circumstances. If a parallel is drawn between these two events, it could be said that the nature of Shi'ism is such that it cannot come to power except by means of insurrection.

39. Ibn Khaldūn, *Muqaddima*, vol. 1, p. 533.

40. Ibid., p. 412.

41. Al-Ghazālī, *Kitāb al-Mustaẓhirī*, pp. 176–177.

42. In the present study the discussion will be limited to the Imāmite doctrine, i.e., the principal branch of Shi'ism, known also as Shi'ism of the Twelve (*shī'a ithnā 'ashariyya*). The other Shī'ī currents such as Zaydism and Isma'ilism will be disregarded since, although they still exist, they do not represent a significant force at the political level.

43. Muḥammad died without leaving a male heir. His only son (Ibrāhīm) departed

this world before the death of the Prophet.

44. A broader analysis of Shi'ism is given by Mehdi Mozaffari, *La conception shi'ite du pouvoir*.

45. On the meaning of this term see G. R. Hawting, "The signification of the slogan *Lā ḥukm illa lillāh* and the references to the Ḥudūd in the Traditions about the Fitna and the Murder of 'Uthmān."

46. The appellation Khārijī is discussed in G. Levi Della Vida in "Khāridjītes," in *Encyclopedia of Islam*.

47. The *shabibiya* current of Khārijism even held that "it was legitimate to confer the imāmate on a woman if it were shown that she were capable of taking the leadership of the community in her hands and of effectively combatting its enemies," Laoust, *Les schismes dans l'islam*, p. 41.

48. *Aḥkām al-sulṭāniyya*, chapter on the emirate. The translation by Léon Ostrorog has been used, with modifications.

49. Ibid.

50. *'Ulamā'* signifies *savants*, men of knowledge. The root of the term is *'ilm* (knowledge and science). Since Muslims believe that a knowledge of their religion is a matter of science, they use the term to refer to all the experts and scholars in Islamic dogma. See Mozaffari, *L'Iran*, pp. 219–32.

51. See Mozaffari, "Les rôles des Ulémā dans les mouvements politiques de l'Iran."

52. Ibn Kathīr, *Al-Bidāya wal-nihāya*, p. 110.

53. "Ihya," II, pp. 124–25, in Laoust, *La politique de Gazalī*, p. 239.

54. See: Carla L. Klausner, *The Seljuk Vezirate: A Study of Civil Administration, 1055–1194*, pp. 22–23.

55. For more information on this constitution see Mozaffari, "Comment ils sont gouvernés," *L'Iran*.

56. R. Khomeini, *Ḥukūmat-é Islami*, p. 46.

57. Ibid., p. 60.

58. Khomeini, *Kashf al-asrār* and *Pour un gouvernement islamique*.

59. For example, on the subject of legislative power, Khomeini says that "In Islam, legislative power comes from religion. The legislature of Islam is the sole source of legislative power. No one has the right to draft a legislation, and no one can carry out any law except the law of *sharī'a*. This is why in Islamic government, the legislative chamber will be replaced by an assembly for planning which will draw up plans for each of the ministries in the light of Islam's prescriptions." *Ḥukūmat-é Islami*, p. 46.

60. E. Mortimer, *Faith and Power*, p. 147.

61. There is no consensus on the date when Ottoman sovereigns began to officially assume the title of Caliph. Some date it back to the reign of Murad I (1319–89) (Halil Inalcik, "The Rise of the Ottoman Empire," in *Cambridge History of Islam: Central Islamic Lands*, p. 320). Others, more prudent, take it back only to the epoch of 'Abd al-Ḥamīd I. The latter signed the treaty he made with Catherine of Russia in 1774 as follows: "The sovereign caliph of the Muhammadan religions" (D. Sourdel in *Encyclopedia of Islam*). Also, G. Lewis observes that the title of "caliph" was not one of those borne by Suleiman the Magnificent. *Turkey*, p. 25.

62. See Gotthard Jäschke, "Mustafa Kemal et la proclamation de la République en Turquie."

63. Kaddafi also came to power without the benefit of the *'Ulāma'*'s assistance. Nevertheless, because of the weakness of nationalism in Libya, he preferred to base his power on Islamism instead of nationalism.

64. In 1928, the Kemalist deputies undertook to revise the constitution decreed on the eve of the proclamation of the republic. This text had two disputed articles, one

stating that Islam was the religion of the Turkish state, the other stating that one of the tasks of the Assembly was to ensure the enforcement of religious laws. These two provisions, manifestly anachronistic, were quite simply suppressed. The Assembly likewise decided to substitute a secular oath for the ancient Islamic formula in the protocol for the investiture of the elected representatives of the people. Thus religion was totally eliminated from the constitution. P. Dumont, *Le poids de l'Islam en Turquie*, p. 92. See also B. Torpak, *Islam and Political Development in Turkey*.

65. The preamble of the constitution of November 22, 1982, elaborated under the auspices of the military regime of General Evern, emphatically proclaims its adherence to the "concept of nationalism, defined by Atatürk, founder of the Republic of Turkey, its immortal leader and hero immortal . . ."

66. Dumont and Torpak.

67. These measures resemble those taken by the Pahlavis in Iran. Reza Shah, who also spoke Turkish, and was profoundly influenced by Atatürk, followed suit, prohibiting the wearing of the *chador* and the turban. His son Mohammad Reza Shah replaced the Muslim calendar by the Persian calendar. His attempt to Latinize the Arab alphabet failed.

68. On this *coup d'état* see: F.-W. Fernau, "Le néo-Kemalisme du Comité d'Unité Nationale . . . ."

69. This party was religious in essence and hence anational. It became flatly antinationalist under the impact of the Khomeini revolution in Iran. During the great demonstration against the regime which Erbakan had organized at Konya just before the military took power, the demonstrators refused to sing the national hymn, preferring religious chants and Islamic slogans. This demonstration provided General Evern with an unexpected pretext for passing to action in the name of Kemalist values.

70. Dumont, p. 106.

71. Khomeini and Erbakan have a long history of friendly relations. Erbakan visited Khomeini during his exile in Turkey in 1963. After the Iranian revolution, Erbakan was received in Iran as a "guest of the state."

72. Ş. Mardin, "Religion and Politics in Modern Turkey, p. 157.

73. The reader is referred to numerous works and articles on the Muslim Brotherhood, for example M. Khadduri, *Political Trends in the Arab World*; K. H. Karpat, ed., *Political and Social Thought in the Contemporary Middle East*; E. Mortimer, *Faith and Power*; A. S. Cudsi, ed., *Islam and Power*; M. Arkoun and L. Gardet, *L'Islam, hier-demain*; A. Abdel-Malek, *La pensée politique arabe contemporaine*; O. Carré and G. Michaud, *Les Frères Musulmans (1928–1982)*.

74. Delanoue, "Une épitre de Ḥasan al-Bannā . . . ," pp. 55 ff.

75. Khomeini, Sermon at the Shaykh Anṣārī mosque on November 14, 1965.

76. Delanoue.

77. In the Arab countries, and in Pakistan, where the influence of the Muslim Brotherhood was considerable, especially on the Jam'iyyat al-'Ulamā'-é-Islami, under the leadership of Maudūdī. The latter reflects the two fundamental aspects of the Islam of the Muslim Brotherhood, namely its universality and its self-sufficiency. On the subject of the Islamic state, Maudūdī writes: "The approach of the Islamic state is universal and total (all-embracing). And the sphere of its activity embraces the whole of human life . . ." (Khurshid Ahmad, *Islam: Its Meaning and Message*).

78. The list of victims of the terrorist acts of the Fedāiyīn is long. It includes three prime ministers (two assassinated: Razmara and Manṣūr; one wounded: 'Ala), two ministers (one assassinated: Hajir, minister of the Court; and one wounded: Faṭemī, Mosaddeq's Minister of Foreign Affairs); the well-known writer and historian Kasravi (assassinated). On the Fedāiyīn-e Islam, see Mehdi Mozaffari, *Les rôles des Ulêma dans les mouvements politiques de l'Iran*.

79. The National Charter with force of law by the presidential decree of July 5, 1976.

80. It must not be forgotten that in the mammoth demonstrations organized against the Shah's regime in Tehran and other Iranian cities the predominant slogan was "May the Islamic Republic be established!" (*Jomhuri-é Islami ijād bāyad gardad!*).

81. S. A. Al-Assiouty, in *Le Maghreb musulman en 1979*, p. 99.

82. Ibid.

83. M. 'Aflaq, "Notre nationalisme libéral face à la discrimination raciale."

84. Ibid.

85. 'Aflaq, "Fi sabīl il-Ba'th."

86. Ibid.

87. Ibid.

88. Ibid.

89. 'Aflaq, "Notre nationalisme libéral . . . ."

90. Ibid.

91. 'Aflaq, "Fi sabīl . . . ."

92. "Parti Ba'th et Nasserisme," cited by Samarbakhsh, *Socialisme en Irak et en Syrie*, p. 29.

93. 'Aflaq, "Fi sabīl . . . ."

94. 'Aflaq, "Notre nationalisme . . . ."

95. Samarbakhsh, *Socialisme en Irak et en Syrie*, pp. 30–31.

96. Mortimer, *Faith and Power*, p. 174.

97. Even the exact figure of the population of Saudi Arabia is unknown.

98. On the 'Unaiza, see the article by J. Tkatsch in *Encyclopedia of Islam*.

99. H. Laoust, *Le précis de droit d'Ibn Qudāma*, p. lv.

100. *Traité de politique juridique pour la réforme du berger et de son troupeau* (*Kitāb al-siyāsa al-shar'iyya fī iṣlāh al-rā'ī wal-ra'iyya*), *Traité de droit public* (*Siyasa shar'iyya*), and *Traité de la tradition* (*Minhāj al-sunna*).

101. Laoust, *Le Traité de droit d'Ibn Taymiyya*, p. xxxv.

102. Ibid., p. xxxiii.

103. The disciples of Āl al-Shaykh prefer "Unitary; Muwaḥiddūn" to the current name Wahhābī; the former is taken from the title of this work, itself inspired by the unity of Allāh. On Wahhābism see: Laoust, *Les Schismes dans l'Islam*, pp. 321–32.

104. Laoust, *Le Traité de droit d'Ibn Taymiyya*, p. xliii.

105. G. Salama, "L'islam en Arabie Saoudite," p. 126.

106. Al-Riyad, 6 Muḥarram 1400 (25 November 1979).

107. Ibn Khaldūn, *Muqaddima* vol. 1, p. 385.

108. Ibid., p. 255.

109. Ibid., p. 277.

110. Ibid., p. 310.

111. The term "Islamic" is absent from the official title of the Jamahiriyya.

112. All of the quotations are from Kaddafi's *Green Book* and discourses, the latter being quoted mainly by S. A. Al-Assiouty, "Le Coran contre le Fiqh en Libye."

113. This comment can apply at most to the Arabs. Kaddafi, himself an Arab, forgets that the Arabs today represent only a numerical minority of Muslims. What should non-Arab Muslims do, who are in the large majority? Are they permitted to make use of *'Ulamā'* interpreters to understand the Koran? The Libyan leader is silent on this question.

114. Mu'ammar Al-Kaddafi, *As-sijjill al-qawmī*.

115. Mozaffari, "La problématique de la République et de l'Islam shi'ite."

116. In opposing in theory the imāmate to the caliphate, the Shī'ī have in practical terms moved away (much further than the Sunnī) from the concept of a republic. The

following points should further clarify the relationship between Shi'ism and this concept.

(a) Shi'ism believes in the nominal and express designation of Muḥammad's successor by Allāh, proclaimed by the Prophet. It rejects any recourse to popular vote.

(b) None of the twelve Shī'ī imāms have been elected to the imāmate. The devolution of this quality takes place by heritage and as a function of kinship ties, generally from the father to the eldest son.

(c) In Shī'ī literature, there is no allusion to the republic *Jomhūr* or *Jumhūriyya* in the positive sense. In classical Shī'ī texts and even in those preceding Khomeini's declaration, the words *Jomhūr* or *'Amma* (in the sense of popular) were always used in an accusatory and pejorative sense, with the finger being pointed at the rival current, Sunnism.

(d) The Shī'ī prince is designated by divine grace (*lutf-ilāhi*), i.e., by providence. Popular vote and consent are therefore excluded as a base of legitimacy.

(e) The term Islamic republic in Iran (*Jomhūri Islam idjād bāyad gardad!*) dates only to the end of 1978 and early 1979. This term was uttered for the first time, acclaimed, and the institution of such a regime demanded, during the huge popular anti-Shah demonstrations in Tehran and other large cities in the country . All the slogans chanted by the demonstrators had been prepared and selected in advance by the organization of the *'Ulamā'* of Tehran, the leaders of which were later to become members of the Council of the Revolution. This organization, in turn, received its directives from the general staff of Khomeini at Neauphle-le-Château.

117. The term *iltiqāṭī* was used for the first time by the Iranian writer Jalal Āl-Ahmad in his pamphlet entitled *Gharbzadégui* (Occidentomania) to illustrate the condition of people who in both their thought and their private lives endeavored to reconcile Western modernity with Islamic traditionalism.

118. The only reference to Allāh is in the quotation of the Koranic formula of *basmala* placed at the head of the text of the program.

119. My translation is from the original text (in Persian).

120. I have stressed the political nature of the fusion between God and the people. This fusion, it must be said, is devoid of any philosophical or theological content. There is, in my view, no relationship between such a political fusion and a fusion that unites the essence of God and man (*wahadat-é wujūd*).

121. On the contradictions between the republic and Islam, see "La problématique de la République et de l'Islam chi'ite."

122. Declaration of July 2, 1981.

123. The Khārijī are cursed by the Shī'ī because they rebelled against 'Alī. In comparing the Mujāhidīn to the Khārijī, Khomeini's regime hoped to discredit them in Shī'ī opinion.

124. Laoust, *Le traité de droit public d'Ibn Taymiyya*.

125. However, this term has been used in the Koran where the Book permits the division of people among diverse tribes, clans, and "nations," vol. 49, p. xlix.

126. Abdel-Malek, *Idéologie et renaissance nationale, l'Egypte moderne*, p. 491.

127. Ibid., p. 489.

128. For more information on this new class see Mozaffari, *La naissance de la bourgeoisie commerçante*.

129. A distinction must be drawn between the term *mellat* and *ra'iyyat*. Sociologically, the first applies to the urban middle classes while the second comprises the common people and in particular the peasantry.

130. For example B. Etienne, "Essai sur le prône politique dans l'Islam contemporain."

131. See Mozaffari, "Le typologie des conflits au Moyen-Orient."

132. A. Cabral, *Palavras de Ordem Gerais*, p. 34.

133. See Mozaffari, *Le régime de la propriété foncière en Iran*.

134. According to Barrington Moore (*Injustice: The Social Bases of Obedience and Revolt*), for the call for rebellion to be obeyed two conditions are necessary: (1) a large number of the oppressed must be persuaded that the proposed change has a chance of succeeding; (2) the groups of the discontented must have concluded that their present situation is the worst possible, and that any other alternative would in any case not be as bad. In the Iranian case, these two conditions were fully met.

Glossary

Ash'arī the school of Orthodox theology founded by Abu' l-Ḥasan al-'Ash'arī (born in 260/873–74) at Baṣra
Ahl al-Bayt family of the Prophet
'amal administrative jurisprudence
amir, *emir* originally a military commander. Used for commander, ruler, or prince
amr authority
'amma the common people
Anṣār the Helpers. Designates the Medinans who supported Muḥammad
'Ashara Mubashshara Council of Ten
'aṣabiyya kinship, *esprit de corps* (term made famous by Ibn Khaldūn)
aṣl-é luṭf principle of grace
aṣl-é taḥkīm principle of arbitration
bay'a the act by which a number of persons, acting individually or collectively, recognize the authority of another person
basmala to pronounce the divine name (also called *tasmiya*)
bāṭinī devotee of esoteric interpretations of sacred texts, particularly associated with Ismā'īlīs
baṭiniyya the sect of *bāṭinī*
Büyük Millet Meclisi Great National Assembly (Turkish)
chador veil
dawla state, dynasty, wealth
ḍa'īf weak
dhu showka powerful
dīn religion
emir see *amir*, *emir*
fatwā opinion on a point of law. The person who gives a *fatwā* is a *muftī*
faqih specialist in Islamic jurisprudence
fiqh Islamic jurisprudence
fitna revolt, sedition, civil war

Fitnat al-Kubrāt the great revolt that led to the assassination of the third caliph, 'Uthmān
fuqahā' pl. of *faqīh*
ghayba occultation
ḥadīth tradition of what the Prophet or an imām said or did
ḥakam arbitrator
ḥākimiyyat sovereignty
ḥaram sacred enclave
hijir anno hejirae, the Islamic dating system
hijra emigration of Muḥammad from Mecca to Medina in 622 A.D.
ḥujjat ul-Islam proof or evidence of the will of Allāh (a title)
ḥukm authority, decree, order
ijmā' consensus
ijtihād independent judgment
al-ijtimā' al-insanī human society
Ikhwān "the Brotherhood," Arab tribesmen belonging to a religious and military movement that had its heyday from 1912–30
iltigāṭī fusionists
imām leader (of prayer or society)
imāmate the office of imām
irāda khalqiyya (or kwaniya) creative will
imārat al-istikfā' emirate by designation
imārat al-istilā' emirate by usurpation
irada tashri'yya legislative will
irth heredity
'iṣma infallibility
istidrāk comprehension
ja'fariyya Imāmiyya, theological school established by Ja'far al-Ṣādiq, the sixth Shī'ī imām
jawhar-é Islam substance of Islam
jihad holy war
jizya tax imposed on persons outside the Umma
kuffār infidels
Lā ḥukma illallāh no decision except with God
Lā ilāha illa Lillāh there is no God but Allāh
al-madinat al-fāḍila the virtuous city
madhhabi person or group who give priority to Islam in their political attitudes (Iranian)
marātib social classes
ma'rūf laws of custom
mashiyya God's will
maṣlaḥa public interest
ma'ṣūm infallible

medina city; specifically, Madinat al-Rasūl (City of the Prophet). Refers to the pre-Islamic, Western Arabian oasis community at Yathrib.
mellat nation
melli nationalist
mufassirūn interpreters
muhājirūn emigrants
mūhibat ilāhi divine trust
Mujāhidīn-é Khalq the Fighting Forces of the People of Iran
mu'jiza miracle
mulk power, possession, kingship
mu'minūm believers
munāfiqūn hypocrites
muslimūn sympathizers
mutā' obeyed
mu'tazilī rationalist movement
naba' the Annunciation
nabī warner, prophet
nabuwat prophecy
nafaqa a tax
naqīb delegate
naṣiha good counsel
naṣṣ an explicit text
nawwāb 'āmm general delegates of the Twelfth Imām
nawwāb khāṣṣ specific delegates of the Twelfth Imām
nāyeb-é imām delegate of the imām
qarya town
qawm tribe
qist fair sharing
qiyās deduction, analogical argument
rasūl emissary, envoy, prophet
Ṣaḥāba Companions
ṣahīh authentic
salafiyya reformist
salṭānat royalty
sha'b nation
shahāda faith (''simultaneous witness and acceptance'')
shar'/sharī'a law of Islam
showka power
shūrā a council
siyāsa policy
siyāsa al-shar'iyya religious policy
sunna what Muḥammad said or did, legal precedent, tradition of the Prophet
ta'aḍud mutual succor

ta'ānus social life
tabaddul change, transformation
tabarrā distance, hostility
ṭabī'a nature
tabi'īn disciples of the Companions
tafaḍḍul preeminence
ṭahāra purification
tanzīl form of revelation
tanẓīmāt reordering/reorganization (term used to describe period of modernization and westernization in the Ottoman Empire, mid-19th century)
taqarrūb close to (somebody)
tariqa method or path
taslīm resignation
tawaḥḥush uncivilized life
tawḥīd oneness of God
tawallā rapprochement, friendship
'Ulāma' those learned in religious knowledge
ukhuwwat fraternity
umma the religious community
umran al'ālām universal civilization
'urf custom, law
waḥy revelation
walāya public function, jurisdiction, the office of government
waṭan fatherland
wāzi' moderator
wazir vizir
walāyat-é faqīh tutelage of *faqīh*
wujūd existence
Yahūd Hebrew tribes
Yathrib the old name of Medina
zaīm leader

Selected Bibliography

Abdel-Malek, A. *Idéologie et renaissance nationale. L'Egypte moderne*. Paris: Anthropos, 1975.

'Aflaq, Michel. "Fi sabīl il-Ba'th." French translation by J.-P. Viennot. *Orient* 29 (1964).

———. "Notre nationalisme libéral face à la discrimination raciale." *Orient* 28 (1963).

Āl-Ahmad, J. *Gharbzadégui* [Occidentomania]. Tehran, 1962.

Ahmad, Kurshid. *Islam: Its Meaning and Message*. London: Islamic Council of Europe, 1976.

Arkoun, M. and L. Gardet. "Autorité et pouvoirs en Islam." In *Pouvoir et verité*. Paris: Cerf, 1981.

———. *Lecture du Coran*. Paris: Maisonneuve et Larose, 1982.

———. *L'Islam, hier-demain*. Paris: Buchet/Chastel, 1978.

———. *Pour une critique de la raison islamique*. Paris: Maisonneuve et Larose, 1984.

Al-Assiouty, S. A. "Le Coran contre le Fiqh en Libye." In *Le Maghreb musulman en 1979*. Paris: CNRS, 1981.

Badie, B. *Culture et politique*. Paris: Economica, 1983.

Al-Baghdādī, Abū Manṣur. *Kitāb uṣūl al-dīn*. Istanbul, 1928.

Brzezinski, Z. *Power and Principle*. New York: Farrar, Straus & Giroux, 1983.

Cabral, A. *Palavras de Ordem Gerais*. Bissau: PAIGC Secretariado General, 1976

Carré, O. and G. Michaud. *Les Frères Musulmans (1928–1982)*. Paris: Gallimard-Julliard, 1983.

Cudsi, A. S., ed. *Islam and Power*. London: Croom Helm, 1981.

Delanoue, G. "Une épitre de Ḥasan al Bannā' aux Frères Musulmans." *Cahiers de linguistique, d'orientalisme et de slavistique* 1–2 (1973).

Dumont, P. "Le poids de l'Islam en Turquie." In *L'Islam et l'État*. Paris: Presses Universitaires de France, 1982.

Etienne, B. "Essai sur le prône politique dans l'Islam contemporain." *Revue française de science politique* (August 1983).

Al-Farabı. *Kitāb Ārā Ahl al-Madīnat al-Fāḍilat*. (*Idées des habitants de la Cité vertueuse*. French trans. by Y. Karam, J. Chlala, and A. Jaussen.) Beirut: UNESCO, 1980.

Fernau, F.-W. "Le néo-kémalisme du comité d'unité nationale." *Orient* 16 (1960).

Al-Ghazālī. *Iḥyā' 'ulūm al-dīn*. Cairo, 1937.

———. *Kitāb al-Mustaẓhirī*. Cairo, 1964.

———. *Naṣihat al-mulūk*. Tehran, 1972.

Hawting, G. R. "The Signification of the Slogan *Lā ḥukm illa lillāh and the Refer-*

ences to the Ḥudūd in the Traditions about the Fitna and the Murder of 'Uthmān.'' BSOAS LXI (1978).
Ibn Kathīr. *Al-Bidāya wal-nihāya*. Cairo, 1932-39.
Ibn Khaldūn. *Discours sur l'Histoire universelle*. 3 vols. Trans. by Vincent Monteil. Paris: Sindbad, 1978.
———. *Muqaddima*. Baghdad, n.d.
Ibn Taymiyya. *Al-Siyāsa al-shar'iyya fī iṣlāh al-rā'ī wal-ra'-iyya*. French trans. *Traité de politique juridique pour la réforme du berger et de son troupeau*. Beirut, 1966.
———. *Minhāj al-sunna*. French trans. *Traité de la tradition*.
Inalcik, Halil. ''The Rise of the Ottoman Empire.'' In *Cambridge History of Islam. Vol. I: Central Islamic Lands*. Edited by P. M. Holt, K. S. A. Lambton, and B. Lewis. Cambridge: Cambridge University Press, 1970.
Al-Isfahānī, Abū 'l-Faraj. *Kitāb al-aghānī*. Būlāg, 1285/1868–69.
Jäshchke, G. ''Mustafa Kemal et la proclamation de la République en Turquie.'' *Orient* 27 (1963).
Al-Kaddafi, Mu'ammar. *The Green Book*. Tripoli, 1981.
———. *As-sijjill al-qawmī*. Tripoli, 1969–79.
Karpat, Kemal H., ed. *Political and Social Thought in the Contemporary Middle East*. New York: Praeger Publishers, 1982.
Khadduri, Majid. *Political Trends in the Arab World: The Role of Ideas and Ideals in Politics*. Johns Hopkins Press, 1970.
Khomeini, R. *Ḥukūmat-é Islami*. Najaf, 1971.
———. *Kashf al-asrār*. Paris: Fayolle, 1979.
———. *Pour un gouvernement islamique*. French trans. by M. Kotobi, B. Simon, and Ozra Bani Sadr. Paris: Fayolle, 1979.
Klausner, Carla L. *The Seljuk Vezirate: A Study of Civil Administration, 1055–1194*. Cambridge: Harvard University Press, 1973.
Laoust, H. *Essai sur les doctrines sociales et politiques de Taki-al-Dīn Ahmad b. Taymiyya*. Cairo, 1939.
———. *La politique de Ġazālī*. Paris: Paul Geuthner, 1970.
———. *Le précis de droit d'Ibn Qudāma*. Beirut, 1950.
———. *Les schismes dans l'Islam*. Paris: Payot, 1965.
———. *Le traité de droit public d'Ibn Taymiyya*. Annotated translation of *Al-Siyāsa al-shar'iyya*. Beirut, 1948.
———. ''Les fondements de l'autorité dans le Minhāj d'al-Ḥillī.'' In *La notion de l'autorité au Moyen Âge*. Paris: Presses Universitaires de France, 1982.
Levi Della Vida, G. ''Khāridjītes.'' In *Encyclopedia of Islam* 2 (new edition). Edited by B. Lewis et al. Leiden: E. J. Brill, 1982.
Lewis, G. *Turkey*. 3rd ed. London: Ernest Benn, 1965.
Makarti, A. M. A. *Water Rights and Irrigation Practices in Lahīj*. Cambridge: Cambridge University Press, 1971.
Makdisi, George. ''Authority in the Islamic Community.'' *La notion d'autorité au Moyen-Âge*. Paris: Presses Universitaires de France, 1982.
Mardin, Ş. ''Religion and Politics in Modern Turkey.'' In *Islam in the Political Process*, edited by James P. Piscatori. Cambridge: Cambridge University Press, 1983.
Al-Māwardī. *Al-aḥkām al-sulṭāniyya*. Beirut: Dar al-Kitāb, 1982.
———. *Les statuts gouvernementaux*. French trans. by E. Fagnan. Algiers, 1915.
———. *Traité de droit public musulman*. French trans. by Léon Ostrorog. Paris, 1901.
Moore, Barrington. *Injustice: The Social Bases of Obedience and Revolt*. New York: M. E. Sharpe, 1979.
Mortimer, Edward. *Faith and Power: The Politics of Islam*. London: Faber & Faber,

1982.
Mozaffari, Mehdi. *La conception chi'ite du pouvoir*. Paris: Sorbonne-Panthéon, 1971.
——. *La naissance de la bourgeoisie commerçante en Iran*. Aarhus: Institute of Political Science, 1981.
——. "La problématique de la République et de l'Islam chi'ite." In *Revue Juridique et Politique*. Paris: EDINA, 1980.
——. *Le régime de la propriété foncière en Iran*. Aarhus: Institute of Political Science, 1981.
——. "Le typologie des conflits au Moyen-Orient." In *Tiers-Monde: diplomatie et stratégie*. Paris: Economica, 1984.
——. *Les rôles des Ulêma dans les mouvements politiques de l'Iran*. Paris: Université de Paris, Faculté de Droit et des Sciences Politiques, 1966.
——. *L'Iran*. Paris: LGDJ, 1978.
Niẓām al-Mulk (Ṭūsī). *Siyāsat-nāmeh*. Paris, 1891.
Salama, G. "L'Islam en Arabie Saoudite." In *Régimes islamiques*. Paris: Presses Universitaires de France, 1980.
Samarbakhsh, A. G. *Socialisme en Irak et en Syrie*. Paris: Anthropos, 1978.
Serjeant, R. B. "The *Sunnah Jāmi'ah*, Pacts with the Yathrib Jews, and the *Taḥrīm* of Yathrib . . ." *BOSOAS* Vol. 41, Part 1 (1978).
Tkatsch, J. "'Unaiza." In *Encyclopedia of Islam* 1. Edited by B. Lewis et al. Leiden: E. J. Brill, 1913.
Torpak, B. *Islam and Political Development in Turkey*. Leiden: E. J. Brill, 1981.

About the Author

Mehdi Mozaffari holds a graduate degree in political science from the University of Tehran, a degree in Arabic from the École des Langues Orientales de Paris, a degree in international relations from the Institut d'Études Politiques de Paris, and a doctorate in political science from the Sorbonne.

Formerly a professor and head of the department of international relations at the University of Tehran, he has also taught at the Institut Universitaire des Hautes Études Internationales de Genève, as well as at the Sorbonne, and has participated in several international congresses. He is currently associate professor and head of the department of international relations at the University of Aarhus, Denmark.

Professor Mozaffari is the author of *Le rôle des Ulêma dans les mouvements politiques de l'Iran* (1966), *La conception chi'ite du Pouvoir* (1971), *Théories des relations internationales*, *La Politique étrangère des pays du Moyen-Orient*, and other books, as well as a great many published articles.